Pre-Algebra

Third Edition

Ideal for Independent practice!

101 Lessons and practices

© Abdulaziz M. Alibarre, 2018

Contents

Welcome Students! 8

Ready for Pre-algebra? 9

1) Integers: Absolute Values 10

2) Ordering Integers: Which is bigger? 11

3) Add Integers-Same Sign 12

4) Add Integers-Different Sign 13

5) Add Integers- More Practice A 14

5) Add Integers- More Practice B 15

6) Subtraction with Integers-1 16

7) Subtracting Integers: The Opposite Sign 17

8) Add or Subtract Integers- More Practice A 18

9) Add or Subtract Integers- More Practice B 19

10) Add or Subtract Integers- Evaluate Expressions-A 20

11) Add or Subtract Integers- Evaluate Expressions-B 21

12) TEST 1: Add or Subtract Integers 22

13) Multiply Signed Integers 23

14) Multiplying Integers- Evaluate Expressions-A 24

15) Evaluate Algebraic Expressions B 25

16) Multiply Algebraic expressions: A 26

17) Multiply Algebraic Expressions-B 27

18) Divide Signed Integers 28

19) Divide Simple Algebraic Expressions 29

20) Evaluate Algebraic Expression: Mixed I 30

21) Evaluate Algebraic Expressions: Mixed II 31

22) Equations: Find the Solution 32

23) Solving Equations: Basic Principles 33

24) Solve One Step Equations: Addition Property 34

25) Solve One Step Equations: Subtraction Property 35

26) Solve One Step Equations: Addition and Subtraction 36

27) One Step Equations: Addition & Subtraction 37

28) Solve One Step Equations: Division Property 38

29) Solve One Step Equations: Division Property II 39

30) Solve One Step Equations: Multiplication Property 40

31) Solve One Step Equations: Multiplication Property II 41

32) One Step: Multiplication/Division Properties 42

33) Solve One Step Equations: Fractions 43

34) Solve One Step Equations: Fractions II 44

35) Solve One Step Equations: Decimals 45

36) Solve One Step Equations: Mixed Review 46

37) TEST II: Solve One Step Equations: Show Work 47

38) Math Language: Translating Math Expressions 48

39) Math Language: Variables 49

40) Equations: Word Problems 50

41) The Distributive Property 51

42) The Distributive Property-More Practice 52

43) Identify like Terms 53

44) Combine like Terms 54

45) Add and Subtract Algebraic Expressions 55

46) Simplify by Combining Like Terms 56

47) Combine (Add/ Subtract) Algebraic Fractions 57

48) Algebraic Fractions= More Practice 58

49) Two Step Equations: Variables on One Side 59

50) Two Step Equations: Variables on One Side II 60

51) Solve Equations: Variables on Both Sides I 61

52) Solve Equations: Variables on Both Sides II 62

53) Solving Proportions Equations 63

54) How to Solve Proportion Word Problems: 64

55) Test III 66

56) Solve and Graph Inequality 67

57) Write the corresponding Inequality 68

58) Exponents: The Basics 69

59) Add or Subtract Exponents 70

60) Multiply and Divide Exponents 71

61) Power of Power (Exponent of an Exponent) 72

62) Negative Exponents 73

63) Evaluate Exponents 74

64) Scientific Notation- from Standard to Scientific 75

65) Scientific Notation- from Standard to Scientific 2 76

66) Scientific Notation-From Scientific to Standard 77

67) Scientific Notation-Mixed 78

68 Review of Perfect Squares 79

69) Square Roots 80

70) Graphing: Plotting Points 81

71) Graphing: Identify the Ordered Pairs 82

72) Graph ordered pairs in the Coordinate System 83

73) PRACTICE: Which coordinate is each person? 84

74) Graphing Linear Equations: How to Find Ordered Pairs 85

75) Graphing Linear Equations: Find Ordered Pairs 86

76) Find Ordered Pairs for Equations with Fractions 87

77) Graphing Linear Equation 88

78) Match the Graphs with their correspondent equations 89

79) Match the Graphs with their correspondent equations-2 90

80) Graph These Linear Equations 91

81) How to Find the Slope from a Graph 92

82) Find the Slope from the Graphs 93

83) Find the Slope from the Graphs 94

84) Find the Slope from Ordered Pairs 95

85) Write the Equation of the Line: Use Slope/Y-intercept: 96

86) Graph the Line from slope and y-intercept 97

87) Find the slope and the Y-intercept from the Graph 98

88) The Pythagorean Theorem 99

89) The Distance Formula 100

90) **POLYGONS IN SNAP SHOT!** 101

91) The Polygons Family Snapshot 102

92) Areas and Perimeters OF POLYGONS (I) 103

93) Circles: AREAS and Circumference 105

94) Area and Perimeter of a kite 107

95) Finding Surface Area and Volume of a Cube 108

96) Finding Surface Area and Volume of Rectangular Prism 109

97) Finding Surface Area of Sphere 110

98) Find Surface Area and Volume of a Cylinder 111

99) Find Surface Area and Volume of a Cone 112

100) Finding Surface Area and Volume of Pyramid 113

101) Review of three Dimensional Area and Volumes 114

Welcome Students!

You did the right choice in deciding to work with this guide. Your time is very valuable and I don't want you to waste it in long exercises and tedious explanations that do not add any meaningful understanding to the concepts and skills required to master pre-algebra.

Instead, this guide is just straight to the point! Each and almost every page starts with quick and short examples or explanations immediately followed by the skill practice. Work on these exercises and check the selected answers at the back of the book! That is it. In no time, you master the pre-algebra concepts and be ready to cruise algebra at high speed!

Congratulations in deciding to master the pre-algebra which is the true entrance to advanced algebra!

1)
$$598 + 72$$ $$305 - 99$$ $$48 \times 72$$ $$708 \times 69$$

2) $8\overline{)128}$ $62\overline{)6918}$ $1\frac{5}{9} + 1\frac{2}{5}$ $7\frac{3}{5} - \frac{2}{7}$

3) $2\frac{2}{9} \times 2\frac{7}{10}$ $\frac{4}{5} \div 3\frac{1}{5}$ $11 - 0.15$ $$5.24 \times 5.8$$

4) $4.2\overline{)84}$ $0.06\overline{)3}$

Fraction	Decimal	Percent
$\frac{3}{15}$	?	?
?	0.04	?
?	?	2.5%

1) Integers: Absolute Values

Integers: Include whole numbers and negative numbers.

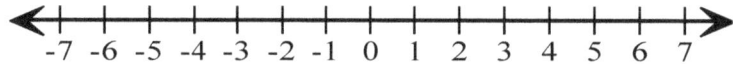

- **Gaining or more of something** is positive on number line.
- **Loosing or less of something is negative** in the number line.
- **The absolute value** of any number is always positive : $|-9| = 9$

Is the integer in each situation positive or negative?

1. The temperature of Minneapolis is 20 degrees below zero.

2. Amina has lost seventy five dollars.

3. Farah has gained a profit of $300.

4. Omar has lost 20 pounds before gaining 30 pounds.

5. A baby grows 9 inches taller.

6. Ali rides the hill for 200 feet.

7. Sofia has removed 10 books from the store.

Evaluate the expressions

8) $|-20|$	9) $|-12|$	10) $|-12| + |-13|$
11) $|-15| + |2|$	12) $-|-17|$	13) $|-12| + |10|$
14) $|-13| + |-6| - 5$	15) $28 - |-17|$	16) $|-12| + |10| - 10$

2) Ordering Integers: Which is bigger?

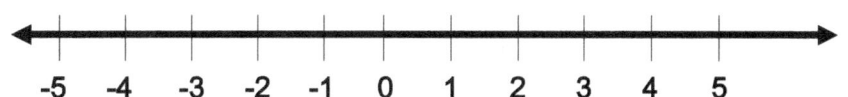

-5 -4 -3 -2 -1 0 1 2 3 4 5

- **A positive number** is always greater than negative number: $4 > -10$
- **For two negative numbers**, the one closer to the zero is greater: $-2 > -5$
- **The absolute value** of any number is always positive: $|-7| = 7$

A) Use either a $<$ or $>$ to say the truth about a number:

1) -8 ◯ -3   5) -2 ◯ -1/2   9) -3/4 ◯ -1/2

2) -3 ◯ -9   6) 0 ◯ -5   10) -4.5 ◯ -5.05

3) 0.5 ◯ -2   7) -17 ◯ -19   11) 2.5 ◯ -7.05

4) $|-7|$ ◯ 20   8) $|-6|$ ◯ -24   12) $|-12|$ ◯ 11

B) Order these numbers from smallest to largest

---

1) -7, -5, 5 ,-9, 0, |-10 |

---

2) -4, -11, 15 ,-91, 0 , $|-15|$

---

3) $|-4|$, -4 , 0, -11, 13, $|-14|$

---

4) $|-14|$, -14 , 1 0, -12, 1, $|-11|$

5) -1, -5, $|-7|$, -4 , -10, -11

6) $|-9|$, -9 , -10, -12, -21

3) Add Integers-Same Sign

To add integers of the same signs: add numbers and keep their common sign.

1) $(-2) + (-6) = -8$ **2)** $-5 + -6 = -11$ **3)** $(5) + (4) = 9$

1) $-6 + (-4)$

11. $-9 + (-14)$

2) $-9 + (-2)$

12. $14 + 9$

3) $-8 + (-4)$

13. $-33 + (-4)$

4) $-1 + 5$

14. $-21 + (-10)$

5) $13 + 4$

15. $-17 + (-4)$

6) $-13 + (-4)$

16. $16 + (5)$

7) $15 + 21$

17. $-20 + (-4)$

8) $-10 + (-14)$

18. $13 + 22$

9) $-8 + (-4)$

19. $(-40) + (-35)$

10) $-14 + (-12)$

20. $-13 + (-4)$

4) Add Integers-Different Sign

To add integers of the different signs: Just **subtract** and choose the **sign of** the integer with the **larger absolute value.**

  1) $(2) + (-6) = -4$ **2)** $-8 + 3 = -5$ **3)** $-5 + 9 = 4$

1. $-10 + (4)$

2. $9 + (-2)$

3. $8 + (-4)$

4. $-7 + 5$

5. $-13 + 4$

6. $13 + (-4)$

7. $-15 + 21$

8. $10 + (-14)$

9. $8 + (-4)$

10. $14 + (-12)$

11. $15 + (-13)$

12. $-14 + 9$

13. $-30 + (4)$

16. $-25 + (10)$

17. $-18 + (4)$

16. $-16 + (5)$

17. $-20 + (-4)$

18. $-13 + 22$

19. $-40 + 40$

20. $-15 + 4$

5) Add Integers– More Practice A

1) $-3 + (-10)$

2) $-15 + (-6)$

3) $15 + (-4)$

4) $-15 + 5$

5) $-1 + 14$

6) $28 + (-4)$

7) $-5 + (-21)$

8) $17 + (-4)$

9) $-12 + (-24)$

10) $-7.5 + (-2.5)$

11) $-14 + 4$

12) $20 + (-14)$

13) $-5 + (-5)$

14) $-10 + 7$

15) $-20 + 6$

16) $-10 + 10$

17) $26 + (-26)$

18) $(-14) + (-5)$

19) $(-10) + (-5)$

20) $(-10.5) + (8.5)$

21) $-14 + 4$

22) $20 + (-14)$

23) $-5 + (-5)$

24) $-10 + 7$

25) $-20 + 6$

5) Add Integers- More Practice B

1) $-11 + 10 + (-1)$

2) $-20 + (-20) + 5$

3) $(-14) + (-5) + 3$

4) $7 + (-10) + (-5)$

5) $-23 + 10 + 3$

6) $-10 + 8 + (-4)$

7) $9 + 15 + (-20)$

8) $-9 + (-9) + 9$

9) $-12 + (20) + 5$

10) $8 + (-9) + 10$

11) $-14 + 4 + (-5)$

12) $20 + (-14) + (-5)$

13) $5 + (-5) + (-7)$

14) $-10 + 7 + (-5)$

15) $-20 + 6 + (-3)$

16) $-10 + 10 + (-8)$

17) $26 + (-26) + (-8)$

18) $0 + (-16) + (-4)$

19) $12 + (-10) + (-9)$

20) $6 + (-4) + (-5)$

6) Subtraction with Integers-1

Subtraction of integers is just like adding integers. See examples below:		
$6 - 4$ $= 6 + (-4)$ $= 2$	$-8 - 5$ $= -8 + (-5)$ $= -13$	$10.7 - 15.8$ $= 10.7 + (-15.8)$ $= -5.1$

1) $-7 - 3$

2) $5 - 2$

3) $3 - 9$

4) $-1 - 5$

5) $-13 - 4$

6) $3 - 5$

7) $-0.17 - 0.7$

8) $-15.1 - 5$

9) $-11 - 11$

10) $-\dfrac{3}{5} - \dfrac{1}{5}$

11) $9 - (-14)$

12) $-14 - 9$

13) $13 - 14$

14) $-20 - 10$

15) $17 - 4$

16) $-15 - 3$

17) $-10.5 - 3.7$

18) $\dfrac{3}{5} - \dfrac{1}{4}$

7) Subtracting Integers: The Opposite Sign

The negative sign (−) means the opposite. It changes:	
the negative sign into Positive:	**the positive sign into negative:**
1) $2 - (-6) = 2 + 6 = 8$	3) $2 - (+6) = 2 - 6 = 2 + (-6) = -4$
2) $-9 - (-5) = -9 + 5 = -4$	4) $10 - (+9.5) = 10 - 9.5 = 0.5$

1. $9 - (-2)$

2. $8 - (-4)$

3. $-7 - (-5)$

4. $-5 - (-7)$

5. $-3 - (-4)$

6. $-15 - (+21)$

7. $10 - (-14)$

8. $-8.3 - (-4)$

9. $14.1 - (-12.5)$

10. $\frac{3}{5} - \left(-\frac{2}{5}\right)$

11. $-14 - (+9)$

12. $-16 - (-4)$

13   $-25 - (10)$

14   $-15 - (-4)$

15. $-16 - (+5)$

16. $-19 - (-4)$

17. $-13 - (-22)$

18. $\frac{3}{4} - \left(-\frac{3}{5}\right)$

8) Add or Subtract Integers– More Practice A

1) $-5 + (-10)$

2) $-12 - 6$

3) $-15 + (-4)$

4) $-5 + 5$

5) $-11 - 14$

6) $-20 - (-3)$

7) $-7 + (-6)$

8) $-25 + (-20)$

9) $-15 - 20$

10) $-14 - 4$

11) $8 + (-14)$

12) $-8 + (-5)$

13) $-15 - 5$

14) $-20 - 6$

15) $-10 + 10$

16) $22 + (-22)$

17) $(-10) - (-3)$

18) $13 - 5$

19) $-11 - 2$

20) $-1 - (-1)$

21) $-15 + (-5)$

22) $-10 - (-7)$

23) $-20 - (-6)$

24) $-3 - (-7)$

25) $-19 - (-5)$

9) Add or Subtract Integers- More Practice B

1) $-5 - 10 + (-1)$

2) $-2 - (-8) + 6$

3) $-4 - 5 - 1$

4) $3 + (-10) - (-2)$

5) $-20 + 9 - 3$

6) $-7 + 8 + (-4)$

7) $8 - (-10) - (-4)$

8) $-7 - (-8) - 9$

9) $14 - 20 + 7$

10) $16 - (-9) + 10$

11) $-10 - 4 + (-5)$

12) $8 + (12) + (-5)$

13) $5 - (-5) + (-10)$

14) $-10 - 17 - (-28)$

15) $-14 + 6 - 13$

16) $-10 - 10 - (-25)$

17) $2 + (-20) + 18$

18) $8 + (-12) + (4)$

19) $6 - (-10) + (9)$

20) $6 + (-4) + (-2)$

10) Add or Subtract Integers– Evaluate Expressions–A

Evaluate Each expression if $x = -1,\quad y = 2,\quad z = -2,$

Example 1: $\quad x + y$
$$= -1 + 2 = 1$$

Example 2: $\quad z - x - y$
$$= -2 - (-1) - 2$$
$$= -2 + 1 - 2 = -3$$

---

1. $-x + y$

2. $-y + x$

3. $-x - y$

4. $-y - x$

5. $x - z + 2$

6. $y + z - 3$

7. $-z + y$

8. $-z - y$

9. $-y - z$

12. $x + y + Z$

13. $x + z - y$

14. $-x + z - y$

15. $-y - z + x$

16. $-y - z - (-x)$

17. $x - y - z$

18. $y - z - (-4)$

19. $10 - x - y - z$

20. $z - y + x + 4$

11) Add or Subtract Integers- Evaluate Expressions-B

Evaluate Each expression if $x = -1,\quad y = 2,\quad z = -2,$

Example 1: $\quad x + y$ $\qquad = -1 + 2 = 1$	**Example 2:** $\quad z - x - y$ $\qquad = -2 - (-1) - 2$ $\qquad = -2 + 1 - 2 = -3$

2) $-y + y$

3) $-x - x$

4) $4 - x + y$

5) $8 - y - y$

6) $x - y - 2$

7) $-y - z - 10$

8) $-z + y - 8$

9) $x - z - y$

10) $10 - y + z$

12) $12 - z - z$

13) $x + z - y$

14) $-x + z - y$

15) $-y - z + x$

16) $-3 - z - x$

17) $-2 + x - y - z$

18) $y - y - (-y)$

19) $10 - x - y - z$

20) $y - z - y - 9$

12) TEST 1: Add or Subtract Integers

Simplify:

1) $-2 + 5$

6) $4 + (-4) + (-5)$

2) $(-5) + 3$

7) $-6 - (-5) + 3$

3) $8 - (-10)$

8) $-10 + 10 - (-5)$

4) $(-1) + (-5)$

9) $-10 - 5 - 5$

5) $-2 + 3$

10) $-8 + 3 - 2$

Evaluate the Expressions if $a = -2$, $b = -1$, and $c = -3$

11) $-a + 8$

16) $-4 - b + c$

12) $a + b$

17) $c - 8 - a$

13) $c - a$

18) $6 + (-a) + (-5)$

14) $a - b - c$

19) $4 - (-c) - (-a)$

15) $10 - c - a$

20) $6 - (-a) + (-b)$

13) Multiply Signed Integers

✓ **To multiply same sign** : Multiply the numbers and keep your answer positive:

      Examples: *1)* $-6 \times -4 = 24$ 2) $(-3)(-5) = 15$ 3) $4 \cdot 5 = 20$

✓ **To multiply different signs:** Multiply numbers & keep your answer negative:

      Examples: *1)* $-7 \times 3 = -21$ 2) $(5)(-2) = -10$ 3) $5(-7) = -35$

Remember *the big dot* (•) and the parenthesis ( ) mean times.

1.	$-3 \cdot (-4)$	**11.**	$9 \times (-1)$
2.	$(-9)(-2)$	**12.**	-1×-9
3.	$8 \times (-4)$	**13.**	$(-3)(-4)$
4.	-1×5	**14.**	$-11 \times (3)$
5.	$-13 \cdot 4$	**15.**	$-7 \cdot -4$
6.	$-13 \times (-4)$	**16.**	$-16(5)$
7.	-15×-2	**17.**	$-8 \times 8 \times 5$
8.	$10 \cdot (-2)$	**18.**	$-3 \times -4 \times -5$
9.	$-1 \times (-4)$	**19.**	$-1 \times -2 \times -5$
10.	9×-4	**20.**	$-10 \times 2 \times -5$

Evaluate Each expression if $x = -2,$ $y = -3,$ $z = 4$

Example 1: $\quad -3x$ $\qquad = -3(-2) = 6$	**Example 2**: $\quad -5xy$ $\qquad = -5(-2)(-3)$ $\qquad = (10)(-3) = -30$

2) $-4x$

3) $-2y$

4) $-3z$

5) $-4yx$

6) **$2xz$**

7) $3xy$

8) $-6yz$

9) $-12z$

10) $-4yz$

12) $-2x$

13) $-3y$

14) $-5y$

15) $-3x$

16) $-3yx$

17) $2xy$

18) $4zx$

19) $-x(-2y)(-z)$

20) $3z(-2y(2x)$

15) Evaluate Algebraic Expressions B

Example: Evaluate Expression if x= -2 and y =3

a) $-6x\,(2y) =$	**b)** $(2)(-4xy)=$
$-6(-2)(2\cdot3) = 12(6)= 72$	$(2)\,(-4\cdot-2\cdot3)= 2(8\cdot3)= 2(24)=48$

Evaluate Each expression if   x = -2 ,   y = 3,   a = 2 and b = -1

1) $-8\bullet(-4x)$

2) $9y\bullet(-3)$

3) $-6\bullet(-4x)$

4) $-9\bullet(-6y)$

5) $6x\div(-3)$

6) $-3(2)(-4xy)$

7) $-4x\bullet(-4)$

8) $-4b(-4a)$

9) $-3a\bullet-4y$

10) $-4a(8x)$

11) $-4xy$

12) $-6y\,(-4)$

13) $-4(8y)$

14) $-3(x)(-y)$

15) $(-8x)(-3x)$

16) $-3a(-4)$

17) $-2a(-4b)$

18) $-2a(-4b)$

19) $(-5x)(-2a)$

20) $(-2x)(-2a)$

16) Multiply Algebraic expressions: A

Multiply the signed numbers and keep the variables:

Examples: 1) $-6 \cdot -2y = 12y$ 2) $3a \cdot (-b) = -3ab$ 3) $-4(5w) = -20w$

1)	$-4\,(-2y)$	**11)**	$-2 \cdot (-4x)(-5y)$
2)	$-2y\,(8)$	**12)**	$-4(-5) \cdot (-2y)$
3)	$-4p \cdot (-5q)$	**13)**	$5m \cdot (-5n)(-1)$
4)	$-3c \cdot (-4d)$	**14)**	$-3(7)(-w)$
5)	$-4a \cdot (-2b)$	**15)**	$-1(-5)(-x)$
6)	$-7r \cdot 8w$	**16)**	$-7\,(0)(-yz)$
7)	$-4(2)(3y)$	**17)**	$-a \cdot (-b) \cdot (-c)$
8)	$-7\,y \cdot (-8z)$	**18)**	$-3 \cdot (-5) \cdot (4z)$
9)	$(-10r)(-2)$	**19)**	$6x\,(-2y(-3z))$
10)	$-6w \cdot (-x)$	**20)**	$-2\,(-4f)(-3)$

17) Multiply Algebraic Expressions-B

Examples: 1) $3 \times 2 = 6$ 2) $-3y \bullet -5 = 15y$ 3) $4y(-5x) = -20xy$

Simplify Each Expression: **Remember** *the big dot* $(\bullet)$ and the parenthesis ( ) mean times.

1. $-6 \bullet (-4)$ 11. $-6 \bullet (-4x)$

2. $6x \times (-3)$ 12. $-4x \bullet (-4)$

3. $-3z \times -4y$ 13. $-4z(0)(xy)$

4. $-4r(8s)$ 14. $(-8)(-2)(-3x)$

5. $-2a(-4b)$ 15. $(-5x)(-2)(3)$

6. $9y \times (-3)$ 16. $-9 \bullet (-6y)$

7. $-3(2) \times (-4xy)$ 17. $-4(-4w)$

8. $-4r \times (8s)$ 18. $-6y(-4)(-2)$

9. $-3(x)(-y)$ 19. $-3w(-4)(2)$

10. $-2(-4c)$ 20. $3(-2)(-2w)$

18) Divide Signed Integers

To divide same sign integers **:** Divide the numbers and keep your **answer positive:**

 Examples: 1) $-6 \div -2 = 3$ 2) $(-20) \div (-5) = 4$ 3) $14 \div 2 = 7$

To divide different sign integers: Divide numbers and keep the **answer negative:**

 Examples: 1) $-16 \div 2 = -8$ 2) $(15) \div (-3) = -5$ 3) $35 \div (-7) = -5$

1. $-8 \div -4$

2. $-20 \div 4$

3. $8 \div (-4)$

4. $(-30) \div (-3)$

5. $(-9) \div (-3)$

6. $-12 \div (-4)$

7. $24 \div -4$

8. $-10 \div 10$

9. $-4 \div (-4)$

10. $-18 \div -2$

11. $9 \div (-1)$

12. $-7 \div -1$

13. $-5 \div 1$

14. $10 \div (-2)$

15. $-9 \div -1$

16. $-16 \div -4$

17. $-4 \div -2$

18. $-8 \div 8 \times 5$

19. $-20 \div -4 \times -5$

20. $7 - 12 \div (-4)$

19) Divide Simple Algebraic Expressions

Example 1: $10 \div 2 = 5$	Example 3: $-15x \div 5 = -3x$
Example 2: $-10x \div -5 = 2x$	Example 4: $15x \div -5 = -3x$

Simplify Each Expression: **Remember** *the big dot* (•) and the parenthesis ( ) mean times.

1. $-6 \div (-2)$ 11. $-16x \div -4$

2. $6x \div (-3)$ 12. $-4x \div (-4)$

3. $-3z \div -3$ 13. $-49z \div -49$

4. $-4r \div (-2)$ 14. $(-8y) \div (-2)$

5. $8b \div (-4)$ 15. $(0) \div (-4y)$

6. $9y \div (-3)$ 16. $-9 \bullet (-6y)$

7. $-12x \div -6$ 17. $-4(-4w)$

8. $-4r \div -2$ 18. $-6y \div (-3)(-2)$

9. $-3x \div (-1)$ 19. $-36w \div (-4) \div (-3)$

10. $12x \div (-1)$ 20. $-12y \div (-2) \div (-6)$

20) Evaluate Algebraic Expression: Mixed I

To evaluate variables, replace the variables with the numbers and simplify:

Examples: **Evaluate Expressions if x= -2, y =3**		
1) $x - 5 =$ $-2 - 5 = $-7	**2)** $-6(x)$ $= -6(-2) = 12$	**3)** $6x \div 2y =$ $6(-2) \div 2\,(3) =$ $-12 \div 6 = -2$

Directions: Evaluate Expressions if x= -2, y =3, a= -3 and b= 4

1.	$6x - 4x$	11.	$19y - 14y$
2.	$-3y - 2y$	12.	$-2a - 6b$
3.	$3x \bullet 4y$	13.	$-3x - 14y$
4.	$-3x \bullet 3x$	14.	$14y - x$
5.	$-2a \bullet 4b$	15.	$-8b \div (-4b)$
6.	$-3y \bullet 4y$	16.	$-16x - (5)$
7.	$13x - 4x$	17.	$-4x + 6y - 2x$
8.	$-6x - (-4y)$	18.	$-6y \div 2y \bullet 5x$
9.	$-8x - (9)$	19.	$-12y - (-5y)$
10.	$6x \div 4y$	20.	$5a - (-4b) + 3y$

21) Evaluate Algebraic Expressions: Mixed II

To evaluate variables, replace the variables with the numbers and simplify:

Example: **Evaluate** $x(y + z)$ when $x = 5$, $y = 3$ and $Z = 1$

$$5(3+1) = 5 \times 4 = 20$$

Evaluate each expression if x = 6, y = 3, and z = 2.

1. $x + 2y + z$ 2. $3x - y$

3. $x + y - z$ **4.** $3x - y + 3z$

5. $12z - x$ **6.** $3(x + y + z)$

7. $xy \div z$ **8.** $xyz - x$

Evaluate each expression if a = 8, b = 4, c = - 6, and d = - 3

9. $a + b - c$ **10.** $a + b - (c + d)$

11. $3a + 4d$ **12.** $bc - d$

13. $(a + b) \div (c - d)$ **14.** $c(4 + d)$

15. $ab - cd$ **16.** $bc + a - d$

> **Example:** Which number is the solution of the equation? $x + 15 = 21$; **4, 5, or 6?**
>
>   The Solution is **6.** Why? Just substitute **x** for **6** to have
>
> $$6 + 15 = 21 \text{ OR } \quad 21 = 21$$
>
>   **Note:** Other numbers such as **4** and **5** wouldn't work. Try them!

1) $x + 15 = 35$; 7, 8, 20

2) $x - 15 = 5$; 10, 15, 20

3) $x + 3 = 25$; 21, 22, 3

4) $x - 7 = 14$; 28, 25, 21

5) $4 + x = 11$; 7, 5, 6

6) $7 + x = 13$; 6,7,8

7) $18 - x = 3$; 14, 15, 16

8) $21 - x = 7$; 14, 15, 18

Find the Solution mentally:

9) $x + 20 = 21$

10) $x - 10 = 21$

11) $x + 17 = 22$

12) $x - 15 = 2$

13) $14 - x = 2$

14) $27 - x = 12$

15) $3 - x = 0$

16) $7 - x = 2$

23) Solving Equations: Basic Principles

➢ **An equation** is a statement that says two expressions or values are equal. Examples are
1) $y + 5 = 8$ **2)** $x + 3 = -2$ 3) $distance = rate \times time$

➢ **Solution of An Equation**: a number you replace with the variable that makes the statement true

Choose the correct solution for $x - 13 = 20;$ $7, 21, 33$

$7 - 13 = 20$	$-5 = 20$	*False*	*therefore, 7 is not a solution*
$21 - 13 = 20$	$8 = 20$	*False*	*therefore, 21 is not a solution*
$33 - 13 = 20$	$20 = 20$	***True***	***therefore, 33 is a solution***

1) $x + 15 = 23; 7, 8, 20$

2) $x - 17 = 3; \ 10, 15, 20$

3) $x - 7 = 5; \ 21, 22, 12$

4) $x + 7 = -1; \ 9, 5, 11$

5) $4 + x = 15; \ 7, 5, 11$

6) $7 + x = -8; \ $ -16,-17, -15

7) $1 - x = -3; \ 4, 5, 6$

8) $30 - x = 5; \ 25, 35, 18$

Find the Solution mentally:

9) $x + 4 = 2$

10) $x - 10 = -3$

11) $x + 10 = 23$

12) $x - 5 = 6$

13) $11 - x = -2$

14) $7 - x = -12$

15) $10 - x = -1$

16) $9 - x = 2$

24) Solve One Step Equations: Addition Property

> ➤ **The Addition Property of Equality:** *Adding the same number to the two sides of an equation doesn't change the solution.*
> **If** $a = b$, **then** $a + c = b + c$
> ***Example***: **If** $5 = 5$ **then** $5 + 3 = 5 + 3$ $8 = 8$ *are still equal*
>
> ➤ This important Property lets you add the same number to the two sides of an equation to solve it!

*Solve for x, y or z using the addition principle: You **MUST** show the work.*

***Example* 1)** $x - 9 = 10$	***Example* 2)** $x - 5 = -15$
Isolate the x by adding 9 to both sides: $$\begin{array}{r} x - 9 = 10 \\ +9 \quad +9 \\ \hline x = \quad 19 \end{array}$$	*Isolate the x by adding 5 to both sides:* $$\begin{array}{r} x - 5 = -15 \\ +5 \quad +5 \\ \hline x = \quad -10 \end{array}$$

1) $x - 4 = 10$ **2)** $x - 2 = -10$ **3)** $x - 3 = -1$ **4)** $x - 7 = -4$

5) $x - 3 = 12$ **6)** $y - 9 = 0$ **7)** $x - 6 = -10$ **8)** $x - 9 = -9$

9) $x - 12 = 10$ **10)** $x - 3 = 10$ **11)** $x - 8 = -1$ **12)** $y - 9 = 11$

13) $x - 7 = -11$ **14)** $z + 1 = -4$ **15)** $x - 8 = -2$ **16)** $x - 2 = -5$

17) $y - 3 - 6 = 11$ **18)** $x + 1 - 2 = -3$ **19)** $z - 1 - 2 = -12$ **20)** $x - 8 = 0$

> ➤ **_The Subtraction Property of Equality:_** _Subtracting the same number from the two sides of an equation doesn't change the solution._
> $$\text{If} \quad a = b, \quad \text{then} \quad a - c = b - c$$
>   **_Example:_** **_If_** $\quad 5 = 5 \quad$ **_then_** $\quad 5 - 3 = 5 - 3 \quad 2 = 2$ **_are still equal_**
>
> ➤ This important Property lets you subtract the same number from the two sides of an equation to solve it!

Solve for x, y or z: You **MUST** *show the work. Follow the example*

1) $x + 15 = 35$	2) $x + 3 = -12$
Isolate the x by subtracting 15 from both sides: $$\begin{array}{r} x + 15 = 35 \\ -15 \quad -15 \\ \hline x = \quad 20 \end{array}$$	*Isolate the x by subtracting 3 from both sides:* $$\begin{array}{r} x + 3 = -12 \\ -3 \quad -3 \\ \hline x = -15 \end{array}$$

1) $x + 4 = 10$ 2) $x + 9 = 12$ 3) $x + 2 = -1$ 4) $x + 7 = -4$

5) $x + 3 = 10$ 6) $y + 8 = 0$ 7) $x + 16 = -10$ 8) $x + 9 = 1$

9) $x + 14 = 14$ 10) $x + 3 = 11$ 11) $x + 8 = -1$ 12) $y + 9 = 11$

13) $x + 7 = -11$ 14) $z + 11 = 4$ 15) $x + 8 = -12$ 16) $x + 9 = -5$

17) $y + 4 - 3 = 11$ 18) $x + 4 + 2 = -10$ 19) $z + 1 + 2 = 12$ 20) $x + 9 = 0$

26) Solve One Step Equations: Addition and Subtraction

*Solve for x, y, or z: You **MUST** show the work. Follow the example*

1) $x - 9 = -10$	2) $x + 5 = -15$
Isolate the x by adding 9 to both sides: $$\begin{array}{r} x - 9 = -10 \\ +9 \quad +9 \\ \hline x = -1 \end{array}$$	*Isolate x by Subtracting five from both sides:* $$\begin{array}{r} x + 5 = -15 \\ -5 \quad -5 \\ \hline x = -20 \end{array}$$

1) $x - 7 = 8$ **2)** $x - 9 = -10$ **3)** $x + 8 = -1$ **4)** $x - 7 = -2$

5) $x + 8 = -12$ **6)** $y - 9 = -5$ **7)** $x + 6 = -13$ **8)** $x + 9 = -9$

9) $x - 12 = 10$ **10)** $x - 3 = 10$ **11)** $x + 9 = -3$ **12)** $y - 9 = -1$

13) $x - 5 = -10$ **14)** $z - 9 = -4$ **15)** $x - 6 = -2$ **16)** $x + 8 = -5$

17) $y - 6 = -13$ **18)** $x + 1 = -4$ **19)** $z - 5 = -15$ **20)** $x + 8 = 0$

Follow the examples to solve for x, y, or z:

1) $-9 + x = -1$
Isolate the x by adding 9 to both sides: $\begin{array}{r} -9 + x = -1 \\ \underline{+9 \qquad +9} \\ x = 8 \end{array}$

2) $-5 = 8 + x$
Isolate the x by subtracting **8** from both sides: $\begin{array}{r} -5 = 8 + x \\ \underline{-8 \quad -8} \\ -13 = x \\ or\ x = -13 \end{array}$

1) $x + 5 = 1$

2) $x - 3 = -6$

3) $3 + x = -7$

4) $4 - 7 = x$

5) $-9 + y = -2$

6) $y + 9 = 6$

7) $y - 6 = -10$

8) $2 + x = -14$

9) $-8 + x = 4$

10) $-13 = x + 4$

11) $-8 = -1 + x$

12) $y - 9 = 11$

13) $x - 7 = -11$

14) $z + 1 = -4$

15) $x + 9 = -13$

16) $x - 3 = -8$

17) $12 + x - 6 = 10$

18) $x - 2 + 5 = -10$

19) $z + 8 - 10 = -1$

20) $x - 15 = 0$

28) Solve One Step Equations: Division Property

➢ **The Division Property of Equality:** *Dividing the same number to the two sides of an equation doesn't change the solution*

$$\text{If } a = b, \quad \text{then } \frac{a}{c} = \frac{b}{c}$$

Example: $\quad 8 = 8, \quad$ *then* $\quad 8 \div 2 = 8 \div 2 \quad 4 = 4$ *are still equal*

➢ This important Property lets you divide the same number to the two sides of an equation to solve it!

See these important four examples:

1. $8x = 16$	$\frac{8x}{8} = \frac{16}{8}$	$x = 2$	
2. $5x = -20$	$\frac{5x}{5} = \frac{-20}{5}$	$x = -4$	

3) $-3x = -1$	$\frac{-3x}{-3} = \frac{-1}{-3}$	$x = \frac{1}{3}$	
4) $-x = 10$	$\frac{-x}{-1} = \frac{10}{-1}$	$x = -10$	

1) $2x = 10$ 　　　　 2) $3x = 9$ 　　　　 3) $2x = -10$ 　　　　 4) $4x = -4$

5) $-2x = -12$ 　　 6) $-6y = -18$ 　　 7) $-3y = -24$ 　　 8) $-7x = -21$

9) $-4x = -14$ 　　 10) $-3x = -1$ 　　 11) $-5x = 10$ 　　 12) $2y = -14$

13) $8x = -16$ 　　 14) $-z = 4$ 　　 15) $-x = -1$ 　　 16) $-x = 5$

17) $-4y = 36$ 　　 18) $-x = 0$

You MUST show your work. Follow the examples:

1. $8 = 4x$ $\dfrac{8}{4} = \dfrac{4x}{4}$ $2 = x$		2) $-2x = -5$ $\dfrac{-2x}{-2} = \dfrac{-5}{-2}$ $x = \dfrac{5}{2}$
Or $x = 2$		3) $-5 = -x$ $\dfrac{-5}{-1} = \dfrac{x}{-1}$ $x = 5$

1) $-2x = -12$ **2**) $-14 = 7x$ **3**) $-x = -12$ **4**) $-1 = -4x$

5) $-4x = 16$ **6**) $-6 = -18x$ **7**) $-2y = 1$ **8**) $-7x = 28$

9) $-4x = -14$ **10**) $-9x = -3$ **11**) $-5 = 10\,x$ **12**) $-2y = -18$

13) $4 = -16x$ **14**) $-z = -9$ **15**) $-x = -5$ **16**) $-x = -8$

17) $-4y = 9$ **18**) $-x = 0$ **19**) $-7y = -14$ **20**) $-3x = 0$

30) Solve One Step Equations: Multiplication Property

➤ **The Multiplication Property of Equality:** *Multiplying the same number to the two sides of an equation doesn't change the solution*

$$\text{If} \quad a = b, \quad \text{then} \quad a \times c = b \times c$$
$$\textit{Example:} \quad 8 = 8, \quad \text{then} \quad 8 \times 2 = 8 \times 2 \quad 16 = 16 \textit{ are still equal}$$

➤ This important Property lets you multiply the same number to the two sides of an equation to solve it!

Example 1: $\dfrac{x}{2} = 13$	**Example 2:** $\dfrac{y}{-5} = -3$
Multiply both sides by 2 $\dfrac{2 \bullet x}{2} = 13 \bullet 2$	Multiply both sides by 5 $\dfrac{-5 \bullet y}{-5} = -3 \bullet -5$
Cancel the 2 with the 5 $\dfrac{\cancel{2} \bullet x}{\cancel{2}} = 13 \bullet 2$	Cancel the -5 with - 5 $\dfrac{\cancel{-5} \bullet y}{\cancel{-5}} = -3 \bullet -5$
Therefore, $x = 26$	*Therefore,* $y = 15$

1) $\dfrac{x}{5} = 2$

2) $\dfrac{x}{4} = -2$

3) $\dfrac{x}{8} = -12$

4) $\dfrac{x}{-5} = 7$

5) $\dfrac{x}{-4} = -9$

6) $\dfrac{x}{-6} = -2$

7) $\dfrac{x}{-3} = -12$

8) $\dfrac{x}{-7} = -7$

9) $\dfrac{x}{7} = -2$

10) $\dfrac{x}{-9} = -7$

11) $\dfrac{-x}{8} = -2$

12) $\dfrac{x}{-12} = 2$

13) $\dfrac{x}{-16} = 2$

14) $\dfrac{x}{-8} = 0$

15) $\dfrac{x}{15} = -8$

16) $\dfrac{x}{9} = -8$

17) $\dfrac{x}{5} = 2$

18) $\dfrac{-x}{3} = 11$

19) $\dfrac{x}{-25} = 0$

20) $\dfrac{x}{-2} = 10$

31) Solve One Step Equations: Multiplication Property II

Example 1: $2 = \dfrac{x}{-3}$	**Example 2:** $\dfrac{y}{-5} = -2$
Multiply both sides by -3 $\quad -3 \bullet 2 = \dfrac{x}{-3} \bullet (-3)$	Multiply both sides by 5 $\quad \dfrac{-5 \bullet y}{-5} = -2 \bullet -5$
Cancel the -3 with the -3 $\quad -3 \bullet 2 = \dfrac{x}{-3} \bullet (-3)$	Cancel the -5 with -5 $\quad \dfrac{-5 \bullet y}{-5} = -2 \bullet -5$
Therefore, $\quad x = -6$	*Therefore,* $\quad y = 10$

1) $\dfrac{x}{5} = -5$

2) $\dfrac{x}{-4} = -5$

3) $\dfrac{x}{8} = -12$

4) $3 = \dfrac{x}{-5}$

5) $\dfrac{x}{-2} = -6$

6) $\dfrac{x}{-6} = 0$

7) $\dfrac{y}{-3} = -1$

8) $-1 = \dfrac{x}{-7}$

9) $\dfrac{-x}{9} = -2$

10) $\dfrac{x}{-9} = 0$

11) $3 = \dfrac{x}{-7} =$

12) $-3 = \dfrac{x}{-10}$

13) $\dfrac{x}{-1} = -2$

14) $\dfrac{x}{-2} = 6$

15) $\dfrac{x}{13} = -4$

16) $\dfrac{x}{-8} = -8$

17) $\dfrac{x}{-15} = -2$

18) $\dfrac{-x}{3} = -1$

19) $\dfrac{-y}{-2} = 0$

20) $\dfrac{-x}{2} = 3$

Follow the examples to solve for x, y, or z:

1) $15 = -5x$	2) $\dfrac{x}{4} = 8$
Isolate the x by dividing -5 to both sides $\dfrac{15}{-5} = \dfrac{-5x}{-5}$ *or* $-3 = x$	Isolate the x by multiplying 4 to both sides: $\dfrac{4 \times x}{4} = 4 \times 8$ *or* $x = 32$

1) $4x = 16$

2) $3x = -1$

3) $\dfrac{x}{-2} = -7$

4) $-5x = 15$

5) $\dfrac{x}{9} = -1$

6) $-3x = -27$

7) $-x = -1$

8) $-5 = \dfrac{x}{5}$

9) $-11 = \dfrac{x}{-5}$

10) $20 = y \div 8$

11) $\dfrac{x}{-7} = 0$

12) $-y = 11$

13) $-\dfrac{x}{8} = -2$

14) $20x = -5$

15) $2x = -10$

16) $\dfrac{x}{-6} = -9$

17) $-x = -51$

18) $27 = -3y$

19) $30x = -5$

20) $-7x = -5$

Follow the examples to solve for x, y, or z:

1) $x + \dfrac{1}{7} = \dfrac{5}{7}$	2) $x - \dfrac{1}{2} = \dfrac{3}{8}$
Isolate the x by subtracting $\frac{1}{7}$ from both sides and then simplify the fraction as needed: $$x + \dfrac{1}{7} - \dfrac{1}{7} = \dfrac{5}{7} - \dfrac{1}{7} \quad or \; x = \dfrac{5-1}{7} = \dfrac{4}{7}$$	*Isolate the x by adding $-\frac{1}{2}$ to both sides and then simplify the fractions:* $$x - \dfrac{1}{2} + \dfrac{1}{2} = \dfrac{3}{8} + \dfrac{1}{2} \quad or \; x = \dfrac{3}{8} + \dfrac{1}{2}$$ $$x = \dfrac{3}{8} + \dfrac{1 \times 4}{2 \times 4} \quad or \; x = \dfrac{7}{8}$$

1) $x + \dfrac{2}{5} = \dfrac{3}{5}$

2) $x + \dfrac{2}{9} = \dfrac{5}{9}$

3) $x + \dfrac{1}{7} = \dfrac{5}{7}$

4) $x - \dfrac{1}{5} = \dfrac{3}{10}$

5) $x - \dfrac{5}{6} = \dfrac{1}{2}$

6) $x + \dfrac{1}{2} = \dfrac{3}{5}$

7) $x + \dfrac{3}{8} = \dfrac{1}{5}$

8) $x + \dfrac{3}{7} = \dfrac{11}{21}$

9) $4 = x - \dfrac{3}{7}$

10) $\dfrac{1}{3} = x - \dfrac{3}{5}$

11) $x + \dfrac{1}{4} = 0$

12) $x + \dfrac{1}{3} = \dfrac{3}{7}$

Simplify: You MUST show your work:

1) $x - \dfrac{2}{3} = \dfrac{1}{3}$

2) $x + \dfrac{2}{7} = \dfrac{5}{7}$

3) $x + \dfrac{1}{8} = \dfrac{5}{8}$

4) $x - \dfrac{1}{4} = \dfrac{3}{8}$

5) $x - \dfrac{2}{3} = \dfrac{1}{2}$

6) $x + \dfrac{3}{10} = \dfrac{3}{5}$

7) $x + \dfrac{3}{8} = \dfrac{1}{2}$

8) $x + \dfrac{3}{14} = \dfrac{1}{7}$

9) $\dfrac{4}{5} = x - \dfrac{3}{7}$

10) $\dfrac{1}{3} = x - \dfrac{3}{8}$

11) $x + \dfrac{1}{4} = 5$

12) $x - 3 = \dfrac{3}{5}$

13) $x - 4 = \dfrac{3}{10}$

14) $x - \dfrac{1}{6} = \dfrac{2}{3}$

15) $x + \dfrac{1}{3} = \dfrac{3}{4}$

16) $\dfrac{1}{3} = x + \dfrac{3}{5}$

17) $x - \dfrac{2}{9} = 0$

18) $x + \dfrac{9}{10} = \dfrac{7}{30}$

35) Solve One Step Equations: Decimals

Follow the examples to solve for x, y, or z:

1) $15 = 0.5x$	**2)** $x - 1.7 = 8.3$
Isolate the x by dividing 0.5 to both sides $\dfrac{15}{0.5} = \dfrac{0.5x}{0.5}$ *or* $30 = x$	Isolate the x by adding 1.7 from both sides: $x - 1.7 + 1.7 = 8.3 + 1.7$ *or* $x = 10$

1) $2x = 1.6$

2) $0.33x = 0.66$

3) $\dfrac{x}{-2} = 0.7$

4) $x - 1.5 = 3$

5) $\dfrac{x}{0.2} = 2.5$

6) $x + 3.7 = -2.8$

7) $0.1\,x = -1$

8) $-0.2 = \dfrac{x}{5}$

9) $-10 = \dfrac{x}{1.2}$

10) $10.5 = y - 2.8$

11) $\dfrac{x}{-7} = 0.1$

12) $-y = 1.9$

13) $\dfrac{x}{0.4} = -0.25$

14) $-2.5x = -12.5$

15) $0.02x = -10$

16) $\dfrac{x}{6} = 0.5$

17) $0.3x = -5.1$

18) $2.7 = 3y$

19) $x + 12.5 = -7.2$

20) $-7x = 3.5$

45

36) Solve One Step Equations: Mixed Review

1) $-3x = 21$

2) $x + 3 = -10$

3) $\dfrac{x}{-3} = -6$

4) $-5x = 5$

5) $\dfrac{x}{8} = -12$

6) $3x = -30$

7) $2 = -10x$

8) $\dfrac{x}{5} = 2$

9) $-13 = \dfrac{x}{5}$

10) $2 = y + 3$

11) $\dfrac{x}{-2} = 0$

12) $-y = -17$

13) $\dfrac{x}{4} = -6$

14) $20 + x = 4$

15) $x + \dfrac{1}{6} = \dfrac{1}{3}$

16) $-2 = \dfrac{x}{-7}$

17) $21 + x = -1$

18) $-4 = -4 + y$

19) $x - \dfrac{7}{9} = \dfrac{1}{3}$

20) $-7x = 35$

21) $5.1 + x = -1.3$

22) $= \dfrac{x}{0.8} = -1.2$

23) $2.8 + x = 11$

24) $-0.5x = 5$

1) $4x = -20$

2) $x - 5 = 0$

3) $\dfrac{x}{4} = -7$

4) $-5x = -15$

5) $\dfrac{x}{-3} = -9$

6) $-9x = -36$

7) $4 = -8x$

8) $\dfrac{x}{5} = 10$

9) $-10 = \dfrac{x}{2}$

10) $-8 = y - 8$

11) $\dfrac{x}{-7} = 0$

12) $-2y = -10$

13) $\dfrac{x}{8} = -2$

14) $-2 + x = -4$

15) $x + \dfrac{1}{4} = \dfrac{1}{3}$

16) $-5 = \dfrac{x}{6}$

17) $10 + y = -11$

18) $-4 = 5 + y$

19) $x - \dfrac{3}{8} = \dfrac{1}{3}$

20) $-x = 35$

21) $3.1 + x = 3.3$

22) $= \dfrac{x}{0.5} = -10$

23) $5.6 + x = 16$

24) $0.4x = 0.8$

38) Math Language: Translating Math Expressions

Addition Words: add, plus, sum, increased, all, total, altogether....	Subtraction Words: subtract, minus, difference, decreased by, reduced, less...
Multiplication: product, times, twice, multiplied by etc.	Division: divide, average, shared evenly or equally, split equally, quotient...

Translate the following math phrases into numerical expressions:

Example: eleven more than twenty $11 + 20$

1. the difference between thirty and fifteen

2. the sum of five, four and three

3. the product of twenty and two

4. fifteen less than 70

5. the quotient of 15 and 7

6. seventeen increased by five

7. eight reduced by seven

8. the quotient of twenty and 5 decreased by four

Translate into verbal expressions:

Example: $11 \times 5 + 20$; the product of eleven and five increased by twenty

9) $5 - 3$	Three less than five or five decreased by 3
10) $7 + 3$	
11) $5/9$	The quotient of five and nine
12) 3×7	
13) $33 \times 2 + 7$	
14) $6 \div 2$	

39) Math Language: Variables

Variables: In math letters such as **x, y, z** etc. are used to represent the unknown variables.

Example: a number increased by 5 is represented as **x + 5**.

Translate each phrase into an algebraic expression. Use letter "x" as the unknown variable. Use also y for the second variable if there is any.

1. eleven more than a number

2. the difference between a number and fifteen

3. twenty pounds more than his weight

4. the product of five and a number

5. fifteen less than the product of five and number

7. the quotient of 7 and a number

8. seventeen increased by ten times a number

9. nine more than the product of three and two

10. the quotient of twenty and 5 decreased by four.

11. twice of her age increased by twice of her age

12. eight times the product of length and width

40) Equations: Word Problems

Example: the sum of 5 and a number is 25. What is the number?
1) **Build the equation:** $5 + x = 25$
2) **Solve:** since $5 + 20 = 25$; x must equal 20

Write the equation and then solve.

1) The sum of a number and 3 is ten.

2) Four times a number increased by 2 is 22.

3) The combined age of Ali and Asha is 27. Ali is 14. How old is Asha?

4) A number increased by 35 is 62.

5) A number decreased by 10 is 42.

6) Five times a number is 45.

7) The quotient of a number and 12 is 4.

8) Since last year, a baby grows 13 inches to become 30 inches. How tall was the baby?

9) The product of a number and 7 is 63

10)   Omar was 150 pounds a week ago, then he lost 3 pounds before gaining more weight to become 180 pounds. How much he gained?

41) The Distributive Property

$$a(b + c) = ab + ac$$

Example 1) $2(3 - 5x) =$	**Example 2**) $-3(2x - 5y) =$
Multiply (distribute) the 2 to both terms: $2 \cdot 3 + 2 \cdot (-5x) = 6 - 10x$	Multiply (distribute) the -3 to both terms : $-3 \cdot 2x + (-3 \cdot -5y) = -6x + 15y$

Use distributive property to write the equivalent algebraic expression:
Remember *the* the parenthesis ( ) mean times.

1. $-6(x - 4)$

2. $6(-3 + x)$

3. $-3(z - 4y)$

4. $-4(r + 8s)$

5. $-2(2a - 4b)$

6. $9(-3 + 2x)$

7. $-3(2x - 4y)$

8. $-4(8 - 3y)$

9. $-3(x - y)$

10. $-5(k - 4m)$

11. $3(5 - 4x)$

12. $-4(2x - 8)$

13. $-4(x + y)$

14. $(-2 - 3x)5$

15. $-5(-2y - 3)$

16. $-9(-6y - 5)$

17. $(7 - 4w)(-4)$

18. $-2(-6y - 4x)$

19. $(x - 4)(-3)$

20. $3(-2 - 2w)$

42) The Distributive Property-More Practice

Example 1) $-5(3x - 4) =$	*Example* 2) $(2x - 3)(-7) =$
$-5 \cdot 3x + (-5) \cdot (-4) = -15x + 20$	$-7 \cdot 2x + (-7) \cdot (-3) = -14x + 21$

1. $2(x - 4)$

2. $5(2 + x)$

3. $-5(d + 4)$

4. $-7(y + 2)$

5. $-3(b - 1)$

6. $2(-3 + 7y)$

7. $-1(-2x - 1)$

8. $(8 - 7y)(-4)$

9. $-10(x - 4)$

10. $-5(-10 - 4)$

11. $(x - 4)(-6)$

12. $(-5w - 10)5$

13. $3(-10q - 3)$

14. $(1 - 4y)6$

15. $-4(2y + 10)$

16. $1(x + 10)$

17. $-1(-5 - 8y)$

18. $-9(-4y - 1)$

19. $-8(-y - 5)$

20. $(7 - 8w)(-3)$

Definition: Like terms or similar terms are exactly the same except for their numerical coefficients.

Examples: Identify like terms

1) $3x, -8x, 4x$ All three are like terms	2) $2x^3, -8x, 4y$ None of them are like terms.
3) $xy, xw, 4y$ None are like terms	4) $2x^2y, -8x^2y, 10x^2y$ All three are like terms.

Identify the like terms if any:

1) $2x, -8y, 4x$

2) $12y, -8, 4x, 12$

3) $2x^2, -8x, -10x^2, 7x$

4) $-4y, 2y, 10y^2$

5) $-x, -y, -7x, 3$

6) $16x^2, 8x, 8, -12$

7) $-4w, 2y, 8y^2$

8) $3x, -1, -4x$

9) $-4y, -4, -4x, -4m, -4y^2$

10) $3m^2, -8x, 4x^2, 7x, 4m^2$

1) $2x + 15 + 5x = 7x + 15$	**2)** $15x - 20x + 5y + 3 - 7$
	$= -5x + 5y - 4$
2x & **5x** are like terms. So add their coefficients (2+5)x =7x	15x & -20x are like terms: $(15 - 20)x = -5x$
15 is by itself.	3 and -7 are like terms = +3-7 = -4
	5y is by itself.

1) $2x + 7x - 21$	**2)** $8x - 2y - 4x$	**3)** $y - 5y$	**4)** $3x - 5x + 2y$
5) $2(-3 + 2x) - 4x$	**6)** $-3x - 3x + 1$	**7)** $-2(-10x - 3) + x$	**8)** $5x - 8x - 2$
9) $-3x + 4x - y$	**10)** $-2y + 8y + 2x$	**11)** $-2(y + 2 - y)$	**12)** $6y - 3 + 9$
13) $2(x - 3) - 4$	**14)** $20x + 8x - 4x$	**15)** $-4y - 2y + 10y$	**16)** $3x - 1 - 4x$
17) $-2(a - 3b) - 4a$	**18)** $16x^2 + 8x - 4$	**19)** $-4y + 2y + 10y^2$	**20)** $-x - 1 - 7x$

45) Add and Subtract Algebraic Expressions

✓ Combine like terms only: (***combine only x and x and not x and y***)

✓ *Remember also: it is usual to write (x) rather than (1x)*

Examples: 1) $4x - 5x = (4 - 5)x = -x$ 2) $-5x + 5x = (5 - 5)x = 0x = 0$

3) $5x - 5y = 5x - 5y$ 4) $8y - 6y = (8 - 6)y = 2y$

1. $-6x - 4x$

2. $-3y - 2y$

3. $8x - (-4y)$

4. $-x - 3x$

5. $-2a - 4b$

6. $3y - (-4y)$

7. $-13m - 4m$

8. $-10x - (-4x)$

9. $-8x - (9)$

10. $6x - (-4x)$

11. $19y - 14y$

12. $-w - 9w$

13. $-3t - 14t$

14. $-20x - x$

15. $-17t - 4t$

16. $-16x - (5)$

17. $-8x + 8x - 5x$

18. $6y - 20y + 5y$

19. $-12y - (-5y)$

20. $25k - (-4k) + 3y$

Follow the examples:

1) $2x + 13x + 5x$	2) $15y - 20y - 3y - 7$
$= (2 + 13 + 5)x = 20x$	$= (15 - 20 - 3)y - 7 = -8y - 7$

1) $4x + 2x + 3x$

2) $8x - 2 - 4x$

3) $y - 5y + 11y$

4) $3a - 5b + 2a$

5) $2y - 3y + 7y$

6) $-3a - 3b + 7a$

7) $(2 + x) + (8 - 4x)$

8) $5(a - b) - 2b$

9) $-8x + 4x + 3x$

10) $-2y + 8y + 2x$

11) $(7b + 2) - (b - 3)$

12) $6y - 6y + 9$

13) $-2(x - 3) + 4x$

14) $5w + 8 - 4w - 9$

15) $-4t - 5t + 10t$

16) $3(x - 1) + 3$

17) $y - 13 + 13$

18) $(x - 2y) + (x + 2y)$

19) $-3y + 3y^2 + 10y^2$

20) $-12t - t - 7t$

Follow the example to simplify the algebraic Fractions:

Example: 1) $\dfrac{x}{3} - \dfrac{x}{5}$	The LCM is **15**
1. Find the LCM to make denominators equal. 3. Add/ Subtract and Simplify	$\dfrac{x}{3} - \dfrac{x}{5} = \dfrac{x \cdot \textcircled{5}}{3 \cdot \textcircled{5}} - \dfrac{x \cdot \textcircled{3}}{5 \cdot \textcircled{3}}$ $= \dfrac{5x}{15} - \dfrac{3x}{15} = \dfrac{2x}{15}$

1. $\dfrac{5x}{2} - \dfrac{x}{2}$

2. $\dfrac{x}{3} - \dfrac{x}{4}$

3. $\dfrac{2x}{3} + \dfrac{x}{9}$

4. $\dfrac{3}{5} - \dfrac{2x}{15}$

5. $\dfrac{4y}{7} - \dfrac{2y}{5}$

6. $\dfrac{3w}{4} - \dfrac{w}{5}$

7. $\dfrac{9t}{7} + \dfrac{2t}{14}$

8. $\dfrac{3k}{4} - \dfrac{2k}{3}$

9. $\dfrac{t}{2} + \dfrac{2t}{5}$

10. $\dfrac{2x}{7} - \dfrac{2x}{5}$

11. $\dfrac{m}{3} - \dfrac{7m}{4}$

12. $\dfrac{3x}{4} - \dfrac{5x}{16}$

13. $\dfrac{t}{3} + \dfrac{t}{3}$

14. $\dfrac{5x}{9} - \dfrac{x}{4}$

15. $\dfrac{n}{7} - \dfrac{n}{7}$

16. $\dfrac{4x}{3} - \dfrac{5x}{4}$

Example: 1) $\dfrac{5y}{8} - \dfrac{3y}{5}$	The LCM is **40**
1. Find the LCM to make denominators equal. *3. Add/ Subtract and Simplify*	$\dfrac{5 \cdot 5y}{5 \cdot 8} - \dfrac{8 \cdot 3y}{8 \cdot 5}$ $= \dfrac{25y}{40} - \dfrac{24y}{40} = \dfrac{1y}{40} = \dfrac{y}{40}$

1. $\dfrac{5y}{3} - \dfrac{y}{4}$

2. $\dfrac{2x}{5} - \dfrac{x}{4}$

3. $\dfrac{2x}{7} + \dfrac{x}{6}$

4. $\dfrac{3x}{8} - \dfrac{2x}{9}$

5. $\dfrac{4y}{5} - \dfrac{2y}{15}$

6. $\dfrac{5w}{8} - \dfrac{2w}{5}$

7. $\dfrac{c}{6} + \dfrac{2c}{5}$

8. $\dfrac{3a}{10} - \dfrac{2a}{20}$

9. $\dfrac{t}{6} + \dfrac{2t}{5}$

10. $\dfrac{x}{7} - \dfrac{2x}{21}$

11. $\dfrac{4m}{5} - \dfrac{7m}{9}$

12. $\dfrac{2x}{5} - \dfrac{5x}{9}$

13. $\dfrac{4t}{5} + \dfrac{2t}{3}$

14. $\dfrac{5c}{7} - \dfrac{3c}{4}$

15. $\dfrac{6v}{14} - \dfrac{n}{7}$

16. $\dfrac{2d}{5} - \dfrac{d}{8}$

49) Two Step Equations: Variables on One Side

1) Subtract 15 from both sides	$2x + 15 = -5$ $\underline{\quad -15 \quad -15 \quad}$ $2x \quad = -20$
2) Divide both Sides by the coefficient of x which is 2	$\dfrac{2x}{2} = \dfrac{-20}{2} \ or \ x = -10$
Check your Solution Correct!	$2(-10) + 15 = -5$ $-5 = -5$

1) $2x + 7 = 21$ **2)** $3x + 2 = -10$ **3)** $2y - 6 = -2$

4) $9x + 8 = -1$ **5)** $3x - 3 = -30$ **6)** $2 = -10x + 12$

7) $-10 = -7x + 4$ **8)** $21 = 8y + 5$ **9)** $2x + 11 - 5 = 8$

10) $15x + 40 = -5$ **11)** $20 + 8x = 4$ **12)** $4x + 5 = -55$

13) $5x - 5 = 5$ **14)** $3y - 13 = 2$ **15)** $-2 = 3x + 1$

50) Two Step Equations: Variables on One Side II

	$\dfrac{x}{4}+8=3$
1) Subtract 8 from both sides	$-8=-8$
	$\dfrac{x}{4}\quad=-5$
2) Multiply both Sides by the denominator which is 4 and simplify	$4\bullet\dfrac{x}{4}=4\bullet(-5)\ or\ x=-20$
Check your Solution	$\dfrac{-20}{4}+8=3$
Correct!	$3=3$

1) $\dfrac{x}{2}+7=-3$

2) $\dfrac{x}{3}+2=-10$

3) $\dfrac{x}{4}-6=-2$

4) $\dfrac{x}{3}+8=-1$

5) $\dfrac{x}{-6}-3=-10$

6) $2=-\dfrac{x}{4}+1$

7) $-10=\dfrac{-x}{2}+4$

8) $10=\dfrac{y}{2}+3$

9) $\dfrac{x}{3}+11-5=8$

10) $\dfrac{x}{5}+12=-5$

11) $-10+\dfrac{x}{-3}=2$

12) $\dfrac{y}{4}-5=-6$

13) $\dfrac{x}{5}-5=5$

14) $\dfrac{y}{5}-1=2$

15) $-2=\dfrac{x}{2}+1$

51) Solve Equations: Variables on Both Sides I

Follow the example to solve equations:

Example: Solve for x:	1) $9x = 3x + 12$	2) $4 + x = 3x + 12$
Step1: Get the variables on one side. Subtract the smaller from both sides. Step 2: Simplify	$9x = 3x + 12$ $-3x \quad -3x$ $6x = 12$ $x = 2$	$4 + x = 3x + 12$ $-x \quad -x$ $4 = 2x + 12$ $4 - (12) = 2x + 12 - (12)$ $-8 = 2x$ $-4 = x$

1) $7x = 2x + 10$

2) $3x = 14 - 4x$

3) $y - 5y = 12$

4) $3x = 5x + 16$

5) $2(-4 + x) = 4x$

6) $-x + 4 = 3x + 12$

7) $-2(5x - 3) = 2x$

8) $5x = 8x - 3$

9) $3x = 4x + 1$

10) $-2y = 8y + 2$

11) $-2(y + 2) = 2y$

12) $6y = 3 + 9y$

13) $2(x - 3) = 4x$

14) $20x = 8x - 24$

15) $4y - 2y = 10y + 8$

16) $3x - 1 = 4x$

17) $-2(a - 3) = 4a$

18) $16 + 8x = 4x$

19) $4y + 2 = 10y + 4$

20) $-x - 1 = 7x$

61

Follow the example to solve equations:

$8x + 10 = -6x + 12$
$8x + (6x) = -6x + (6x) + 14$
$14x = 14$
$x = 1$

$10 - x + 5 = -8x - 6$
$15 - x + (8x) = -8x + (8x) - 6$
$15 + 7x = -6$
$15 - 15 + 7x = -6 - 15$
$7x = -21$
$x = -3$

1) $4x - 5 = 2x + 10$

2) $7 - 3x = 11 - 2x$

3) $8y - 3y = 2y + 6$

4) $2 - 4(2 + x) = 14$

5) $-10x + 6 = 3x + 15$

6) $-16 + 4x = 2x + 8$

7) $-15 - 6x = -4x + 3x$

8) $9 - 4y = 7y - 13$

9) $-2(-x - 8) = -2x$

10) $2 - 5(x - 3) = -4x$

11) $20x - 6x = -4x - 24$

12) $10 - 4y - 2y = 4y$

13) $-20(a - 3) = -4a - 12$

14) $16 - 8x = -4x - 4$

1) ⇨ A proportion represents two ratios (fractions) that are equal:

2) **To solve a proportion:** multiply the diagonal numbers or variables & solve:

$\dfrac{2+x}{5} = \dfrac{6}{?}$	$\dfrac{y}{5} = \dfrac{9}{7}$	$\dfrac{9}{2} = \dfrac{3p}{6}$
$2(2+x) = 30$ $4 + 2x = 30$ $x = 13$	$7y = 5 \times 9$ $y = 45 \div 7$ $y = \dfrac{45}{7}$	$6p = 54$ $p = 54 \div 6$ $p = 9$

1) $\dfrac{x}{4} = \dfrac{7}{2}$

2) $\dfrac{3x}{8} = \dfrac{7}{2}$

3) $\dfrac{x+4}{5} = \dfrac{6}{2}$

4) $\dfrac{2x+2}{6} = \dfrac{6}{7}$

5) $\dfrac{4}{x+3} = \dfrac{6}{5}$

6) $\dfrac{4}{4+y} = \dfrac{2}{3}$

7) $\dfrac{9}{4} = \dfrac{y+7}{2}$

8) $\dfrac{5}{4} = \dfrac{y+3}{2}$

9) $\dfrac{9}{5} = \dfrac{y+5}{2}$

10) $\dfrac{3}{4} = \dfrac{12}{y+3}$

11) $\dfrac{1}{5} = \dfrac{21}{5y}$

12) $\dfrac{1}{4} = \dfrac{7}{2y-4}$

13) $\dfrac{t}{0.4} = \dfrac{1+t}{0.8}$

14) $\dfrac{4}{5} = \dfrac{9+n}{n-6}$

15) $\dfrac{8x-2}{0.6} = \dfrac{8}{0.4}$

Example 1: Hafsa runs each morning 4 miles in 15 minutes. How many minutes will it take her to run 20 miles?

Step1: Use variable, like x, d etc. for the missing number and set up the proportion (4 miles in 15 Minutes means divide).	$\dfrac{4 \text{ miles}}{15 \text{ minutes}} = \dfrac{20 \text{ miles}}{x \text{ minutes}}$
Step 2: **Cross Multiply:**	$15 \times 20 = 4x$
Step3: **Solve the proportion** (Isolate variable x)	$\dfrac{15 \times 20}{4} = \dfrac{4x}{4}$ $x = 75 \; minutes$

We can exchange denominators and numerators as far as the corresponding units are in the same side). Try this and call me if you do not get the same answer as before!)

$$\frac{15 \text{ minutes}}{4 \text{ miles}} = \frac{x \text{ minutes}}{20 \text{ miles}}$$

Example 2:

In a scale drawing 2 inches represent 30 miles. What distance does a line segment of 7 inches represent?

Step1: Set up the proportion (Let's call d the distance)	$\dfrac{2 \text{ inches}}{30 \text{ miles}} = \dfrac{7 \text{ inches}}{d}$
Step 2: Cross Multiply:	$2d = 7 \times 30$
Step3: Solve the proportion (Isolate the d)	$\dfrac{2d}{2} = \dfrac{210}{2}$ $d = 105 \; miles$

Solve the following Proportion Problems:

1) There was just 2 computers shared by every 5 students. If there were 30 students in class, how many computers they share?

2) Seven students out of every 10 Pre-university students who took the math test passed the test. This month 80 will take the test. How many are expected to pass the test.

3) It takes 2 cubs of sugar to prepare 50 cakes. How many cakes that can be prepared from five and half cubs of Sugar?

4) On a typical school day 5 students in every 75 students claim to have lost their pencils. How many students that will claim to have lost their pencils in a school of 240 students?

5) Gloria runs 6 miles in 30 minutes. At that rate, how far could she run in 120 minutes?

6) Assume test grades are proportional to study time. Asma has studied 4 hours per week to score 80. How many hours should Asma study math per week if she wants to get 98?

7) Waris is preparing cake. She knows that every six cups of flour needs one cup of sugar. If she wants to use 36 cups of flour, how many cups of sugar she should have used?

8) Jack drove his Honda for 200 miles and used 5 gallons of fuel. How long he traveled if he used 18 gallons?

Follow the example to solve equations:

$$2(6x + 4) = 4(x - 3) - 4$$

Distribute	$12x + 8 = 4x - 12 - 4$
Regroup Variables	$8x = -12 - 4 - 8$
Simplify	$8x = -24$
	$x = -3$

1) $4(x - 1) = 20$

2) $5(x + 2) = 40$

3) $8(y + 2) = 2(y + 5)$

4) $-2(2 + y) = 14y + 12$

5) $-2x + 6(x + 1) = 3x$

6) $7(x - 2) = 2(x + 3)$

7) $-14 + 6(x - 1) = -4x$

8) $2k - 2 = -2(k - 7)$

9) $2x = 5(x - 3)$

10) $-5(x - 3) = -4x + 2$

11) $\dfrac{5x}{4} = \dfrac{x + 3}{2}$

12) $\dfrac{y - 3}{5} = \dfrac{y - 2}{4}$

56) Solve and Graph Inequality

Solve Inequality just like you do in the equality but use the inequality sign:

Less <,   **Less or equal ≤,**   **Greater >, Greater or Equal ≥**

See these examples:

---

1) $x + 3 \geq 1$ $x + 3 + (-3) \geq 1 + (-3)$
                     $Or \ \ x \geq -2$

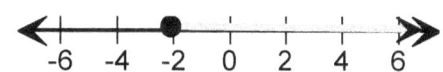

                     See the interval is closed (dark)

2) $-4 < x \leq 2$ See how one side is open and the other side is closed

3) $-2x \leq -8$ $\dfrac{-2x}{-2} \geq \dfrac{-8}{-2}$ $or \ \ x \geq 4$

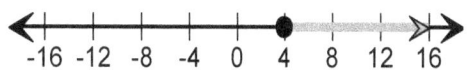

                     Note how the inequality **changes direction** when multiplied or divided by negatives.

Solve (if needed), then graph the inequality

1) $x \geq 1$

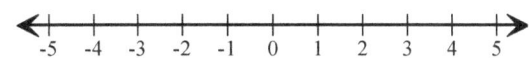

2) $y - 5y \leq 12$

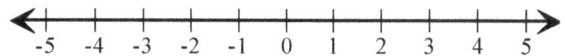

3) $-3 < x \leq 1$

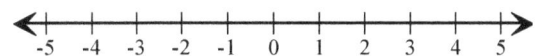

4) $5x < 8x - 3$

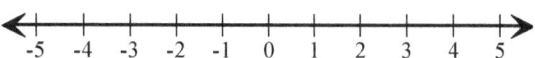

5) $3x + 6 > 2x + 1$

6) $-2 \leq x \leq 1$

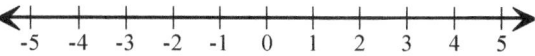

57) Write the corresponding Inequality

1) $x > 2$

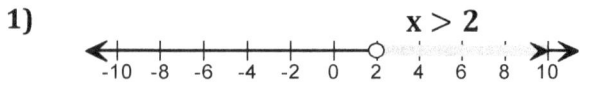

2)

3)

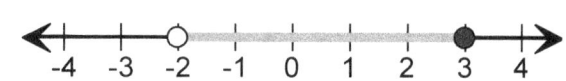

4)

5)

6)

7)

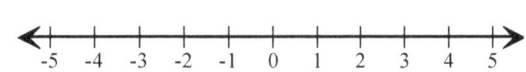

8)

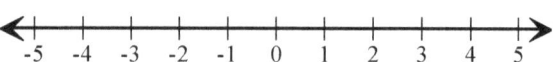

9)

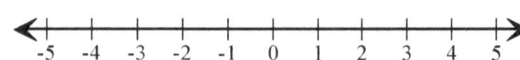

10)

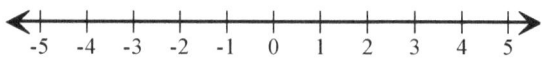

Solve and draw the inequality

11) $x + 5 \geq 1$

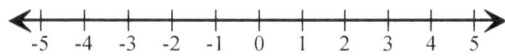

12) $4x \leq 8 + 2x$

13) $-2w - 4 \leq 4$

14) $3x - 12 < 3$

15) $-3x - 6x > -18$

16) $-2y + 6 < -2$

1. **An exponent represents repeated multiplications:**

   - ✓ Instead of writing: $2 \times 2 \times 2 \times 2$, Simply write: 2^4
   - ✓ Instead of writing: $y \cdot y \cdot y \cdot y \cdot y$, Simply write y^5

2. **The base and the power:**

   - ✓ The **2** and **y** are called bases
   - ✓ The **4 & 5** are the exponents or powers

3. **Read it :**

   - ✓ 5^3 as, five to the third power
   - ✓ y^4 as, y to the fourth power

4. **Zero power:**

   - ✓ *Any number that has **zero** as an exponent is 1*
   - ✓ *Therefore,* $(-127)^0 = 1$ *;* $y^0 = 1$ *;* $any\ number^0 = 1$

5. **To add or subtract variables:**
   - ✓ Add only same variables with the same power. $3x^3 + 2x^3 = 5x^3$
   - ✓ Never add a variable and a number: $2x^2 + 5 = 2x^2 + 5$

6. **Never add two powers:**

   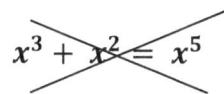

Exercise: *Simplify*

1) $9x^2 + 4x^2 =$

2) $5y^2 - 3y^2 + y =$

3) $2x^3 + x^0 + 5 =$

4) $3^3 + 2x^3 + 5x^3 =$

5) $10y^2 - 3y^2 + 2x^2 =$

6) $7^3 - 5x^2 + x =$

59) Add or Subtract Exponents

Example 1: $3^4 + x^2$ $= 3 \times 3 \times 3 \times 3 + x^2$ $= 81 + x^2$	Example 2: $5^2 + 7^2$ $= 5 \times 5 + 7 \times 7$ $= 25 + 49$ $= 74$	Example 3: $3x^2 + 6x^2 =$ $(3 + 6)x^2 = 9x^2$

(1) $10^0 =$

(2) $3^0 + 1^0 + x^0 =$

(3) $5^3 - 1^0 =$

(4) $3^4 - 3^0 =$

(5) $5^2 - 6 =$

(6) $9x^2 + 7x^2 =$

(7) $5y^7 + 4y^7 =$

(8) $0^5 + 2^2 =$

(9) $3^5 + 2^2 =$

(10) $5y + 3y^2 - 2y =$

(11) $0^5 + 10\,y^4 - 7y^4 =$

(12) $4^4 + 2^2 + 17^0 =$

(13) $10y^4 - 17y^4 =$

(14) $b^5 + 10\,y^4 - b^5 =$

(15) $4y^3 + 2y^2 + 17y^2 =$

60) Multiply and Divide Exponents

Simplify the Exponents. Follow the examples:

$2^3 \times 2^2$ ➡ $2^{3+2} = 2^5$	$x^5 (x^3)$ ➡ $x^{5+3} = x^8$
$\dfrac{3^7}{3^2} =$ ➡ $3^{7-2} = 3^5$	$\dfrac{y^8}{y^3} =$ ➡ $y^{8-3} = y^5$

(1) $3^4 \times 3^4 =$

(2) $x^4 \times x^8 =$

(3) $\dfrac{7^3}{7} =$

(4) $10^0 \times 3^4 =$

(5) $12^0 \times 12^3 =$

(6) $\dfrac{3^7}{3^2} =$

(7) $y^0 \times y^4 =$

(8) $\dfrac{m^7}{m^3} =$

(9) $5^2 - \dfrac{10^2}{10} =$

(10) $5 \times 3^2 - 27^0 =$

(11) $\dfrac{3^7}{3^2} - 3^5 =$

(12) $\dfrac{x^7}{4^3} \times \dfrac{x^7}{4^2} =$

61) Power of Power (Exponent of an Exponent)

Simplify the Exponents. Follow the examples:

$$(x^3)^2 = \Rightarrow x^{3 \times 2} = x^6$$

$$(y^4)^2 = \Rightarrow y^{4 \times 2} = y^8$$

$$x^5(x^3)^2 \Rightarrow x^5(x^6) = x^{11}$$

$$\frac{y^{12}}{(y^5)^2} = \Rightarrow y^{12-10} = y^2$$

(1) $(x^3)^3 =$

(2) $(x^5)^2 =$

(3) $(y^3)^2 =$

(4) $(x^5)^0 =$

(5) $(y^0)^2 =$

(6) $(2^3)^2 =$

(7) $\dfrac{y^6}{(y^2)^2} =$

(8) $3^2(2^2)^2 =$

(9) $x^5(x^5)^2 =$

(10) $2^0(2^4)^2 =$

(11) $(3^2)^2 =$

(12) $\dfrac{y^{12}}{(y^5)^0} =$

62) Negative Exponents

Simplify the Exponents and change all **negative exponents** into positive.
Follow the examples:

$$y^{-4} = \frac{1}{y^4}$$

$$2^{-3} = \frac{1}{2^3} = \frac{1}{8}$$

$$y^{-4} \times y^7 = y^{-4+7} = y^3$$

$$\frac{x^{-5}}{x^2} = \frac{1}{x^2} \times \frac{1}{x^5} = \frac{1}{x^7}$$

(1) $x^{-3} =$

(2) $(x^{-5})^2 =$

(3) $(y^{-3})^2 =$

(4) $\frac{1}{y^{-9}} =$

(5) $(y^{-4})^0 =$

(6) $(2^{-3})^{-2} =$

(7) $\frac{x^{-5}}{x^{-2}} =$

(8) $\frac{x^{-5}}{x^{-2}} =$

(9) $\frac{x^{-7}}{x^{-2}} =$

(10) $\frac{x^9}{x^{-5}} =$

(11) $\frac{x^{-5}}{x^{-5}} =$

(12) $\frac{x^{-5}}{x^{-6}} =$

Evaluate the following exponents. Follow the examples:

$x^3 \times 3^2$ if $x = 2$
$= (2)^3 \times 3^2$ $= 2 \times 2 \times 2 + 3 \times 3$ $= 8 + 9$ $= 17$

$\dfrac{y^7}{y^4}$ when $y = 3$
$y^{7-4} = y^3$ $= 3^3$ $= 27$

(1) $x^4 = $ _____
      if x = 3

(2) $\dfrac{m}{7} = $ _____
      if $m = 21$

(3) $m^5 \times 2 = $ _____
      if m=2

(4) $\dfrac{x^7}{3^2} = $ _____
      when x=1

(5) $y^0 \times y^4 = $ _____
      when y =4

(6) $5^2 \times \dfrac{10^2}{10} = $ _____

(7) $n + n^2 - 8 = $ _____
      When n = 4

(8) $\dfrac{x^6}{x^5} \times \dfrac{x^3}{x^2} = $ _____
      When x = 4

(9) $\dfrac{m^5}{m^2} - 81^0 = $ _____
      If m=9

10) $\dfrac{n^4}{n^2} + n^2 = $ _____
      If n=5

64) Scientific Notation– from Standard to Scientific

✓ The weight of an atom is very small: 0.0000000000000000000000000009109 g

✓ The distance to the sun is very large: 9,296,000,000 miles

It was not easy for scientists to write often these small and large numbers all the time. So, they came up with the idea of writing these numbers in a short form called **Scientific Notation. So, the examples above in scientific notation would be written as:**

Standard Notation (the normal way)	Scientific Notation
0.0000000000000000000000000009109 g	$9.109 \times 10^{-28} g$
9,296,000,000 miles	$9.296 \times 10^{-9} g$ miles

Rules for changing **Small numbers** to scientific notations: Follow the examples:

Steps	Standard Notation	Scientific Notation
✓ First, put a decimal next to the first number at the right of zeros. ✓ Since the decimal has moved four places to the right make the exponent -4.	1) 0.**000147** 2) 0.**000831**	1.47×10^{-4} 8.31×10^{-4}

Standard	Scientific	Standard	Scientific
1) 0.00041		5) 0.0025	
2) 0.0045		6) 0.0000051	
3) 0.000091		7) 0.000805	
4) 0.00000075		8) 0.00045	

65) Scientific Notation– from Standard to Scientific 2

Express each of these **big numbers** number in scientific notation. See examples below:

	Examples
1. First, put a decimal next to the first number from left.	$3521.8 = 3.5218 \times 10^3$
2. Since the decimal has moved 3 places to the left make the exponent 3.	$2643 = 2.643 \times 10^3$
3. Remember the decimal for 2643 is hidden at the end. So, it is indeed 2643.0	

Standard	*Scientific*
1. 45	
3. 41,000	
5. 21	
7. 610,000	
9. 302	
11. 210	
13. 136,000	
15. 81,000	

Standard	*Scientific*
2. 4000	
4. 310	
6. 1000	
8. 45	
10. 21000	
12. 4100	
14. 410	
16. 56	

66) Scientific Notation–From Scientific to Standard

✓ Since the exponent (4) is positive, move the decimal point 4 times to the right. Add zero if you need it.	1) $3.52 \times 10^4 = 35200$ 2) $2.643 \times 10^4 = 26430$
✓ Since the exponents is negative move the decimal point 3 times to the left and 2 times to the left for the second example.	3) $1.47 \times 10^{-3} = 0.00147$ 4) $8.31 \times 10^{-2} = 0.0831$

1. 3.52×10^3 2. 2.8×10^2

3. 5.61×10^{-4} 4. 9.63×10^{-3}

5. 8.02×10^3 6. 7.12×10^4

7. 4.6×10^{-1} 8. 3×10^{-5}

9. 6.5×10^5 10) 5.22×10^6

11) 8.213×10^6 12) 1.61×10^{-5}

13) 3.5×10^{-6} 14) 9.13×10^0

15) 4.21×10^0 16) 3.0×10^1

67) Scientific Notation-Mixed

Fill the missing scientific or Standard Notation

Scientific	Standard		Scientific	Standard
5.3×10^5				0.00033
6.1×10^{-2}				0.4
	0.034		4.152×10^7	
	1100		3.5×10^4	
3.85×10^{-3}				0.0004.5
8.3×10^3				4100
	0.00038		4500.6	
	18,000		3.52×10^4	
1.95×10^{-4}				4100
8.43×10^{-5}				610,000
	810,000		3.67×10^0	
	6,140,000		5.256×10^8	
9.8×10^{-1}				15,000
6.68×10^6				413

A **number squared** means a number multiplied by itself. Here are some examples:

$$5^2 = 5 \cdot 5 = 25 \qquad 7^2 = 7 \cdot 7 = 49 \qquad 10^2 = 10 \cdot 10 = 100$$

To get stronger in mental math, do and memorize these perfect squares.

Exercise 1: Simplify the following squares;

$0^2 =$	$1^2 =$	$2^2 =$	$3^2 =$
$4^2 =$	$5^2 =$	$6^2 =$	$7^2 =$
$8^2 =$	$9^2 =$	$10^2 =$	$11^2 =$
$12^2 =$	$13^2 =$	$14^2 =$	$15^2 =$
$16^2 =$	$17^2 = 289$	$18^2 =$	$19^2 =$
$20^2 =$	$21^2 = 441$	$22^2 =$	$23^2 = 529$
$24^2 =$	$25^2 =$	$26^2 = 676$	$27^2 =$
$28^2 =$	$29^2 = 841$	$30^2 =$	

Square roots are the reverse of the squares.

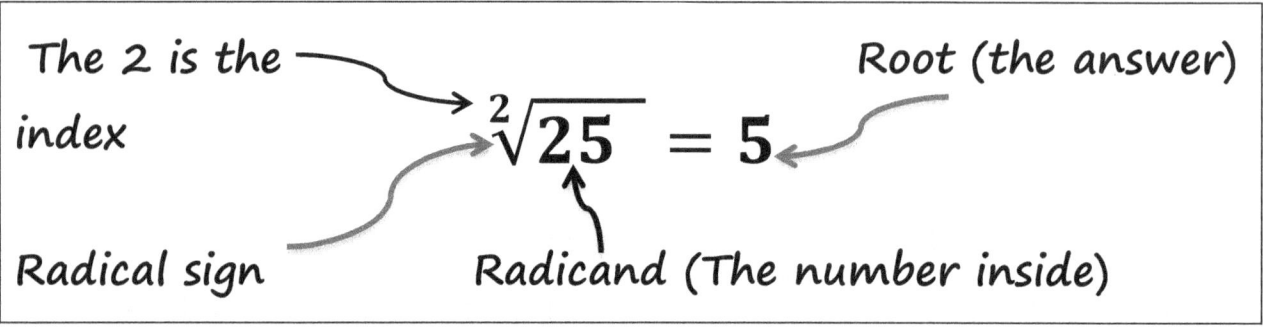

The 2 is the index

Radical sign

$\sqrt[2]{25} = 5$

Root (the answer)

Radicand (The number inside)

Since $5^2 = 25$, the square root of 25 is 5

Note: The square root is so common that we usually omit the indices or the 2 and write simply it as: $\sqrt{25}$

Find the square roots of the following: Refer back to the perfect squares if you need to:

$\sqrt{16}$	$\sqrt{49}$	$\sqrt{1}$	$\sqrt{36}$
$\sqrt{196}$	$\sqrt{100}$	$\sqrt{225}$	$\sqrt{289}$
$\sqrt{9}$	$\sqrt{4}$	$\sqrt{81}$	$\sqrt{64}$
$\sqrt{400}$	$\sqrt{841}$	$\sqrt{169}$	$\sqrt{441}$
$\sqrt{81}$	$\sqrt{256}$	$\sqrt{529}$	$\sqrt{121}$

70) Graphing: Plotting Points

Do you see from both the table and the graph how many minutes are exercised when 12 calories were burned? In math, it is common to use the **ordered pairs** as below:

$$(1, 2), (2, 4), (3, 6), (4, 8), (5, 10), (5, 10)$$

x	y
Minutes in Stair Climber	Calories Burned
1	2
2	4
3	6
4	8
5	10
6	12

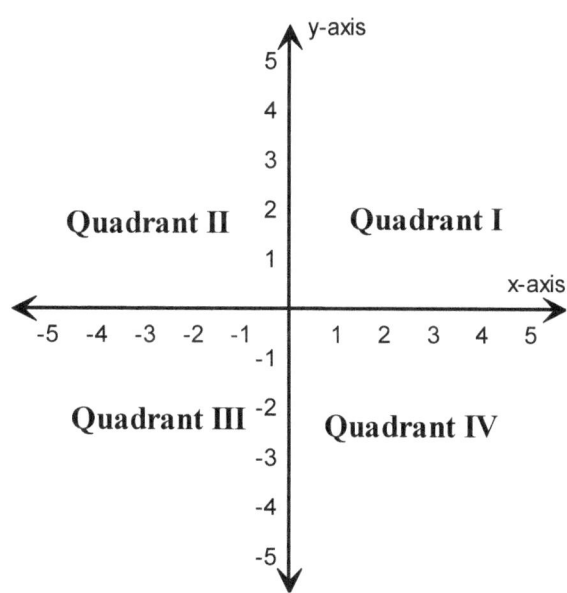

- ✓ The first number of the ordered pair is the **x-coordinate**. The second number of the ordered pair is called the **Y-coordinate.**
- ✓ **The Coordinate System**: sometimes also called the Cartesian System is the system that puts ordered pairs in the graph.
- ✓ *The x-axis and y-axis* divide the system into four regions called Quadrants;
- ✓ **The Origin**: is where the x and y-axis meet and is zero: **(0,0)**

81

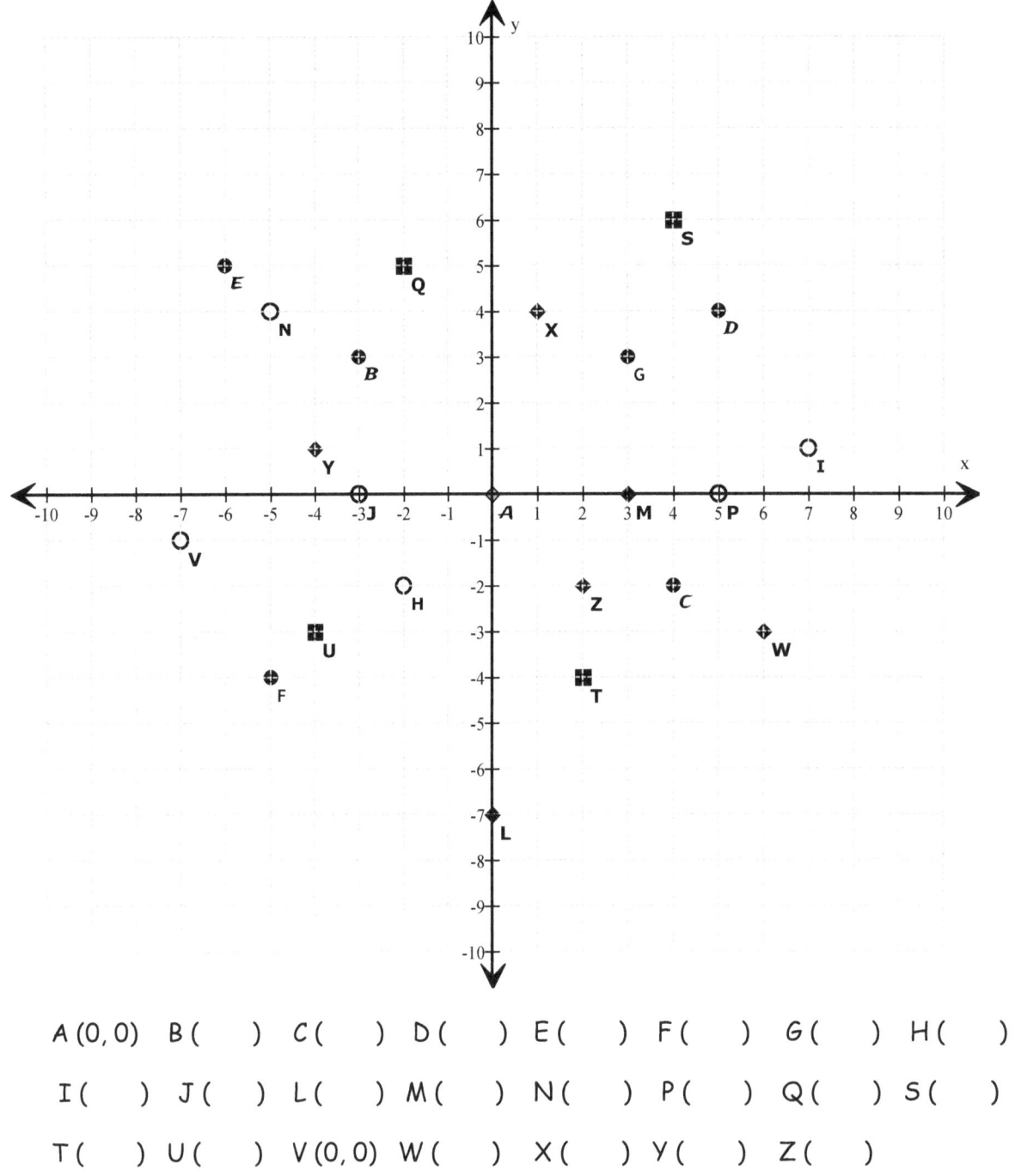

A (0,0)   B (   )   C (   )   D (   )   E (   )   F (   )   G (   )   H (   )

I (   )   J (   )   L (   )   M (   )   N (   )   P (   )   Q (   )   S (   )

T (   )   U (   )   V (0,0)   W (   )   X (   )   Y (   )   Z (   )

72) Graph ordered pairs in the Coordinate System

1. A (3, 4)

2. B (3, -2)

3. C (3, 1)

4. D (2, 4)

5. E (-3, 4)

6. F (-2, 3)

7. G (-5, 1)

8. H (0, 4)

9. I (3, 0)

10. J (-3, -4)

11. K (-1, -2)

12. L (-4, 4)

13. M (-6, 4)

14. N(-5, 3)

15. O (-5, 4)

16. P (3, 4)

17. Q (3, 4)

18. R (-3, -4)

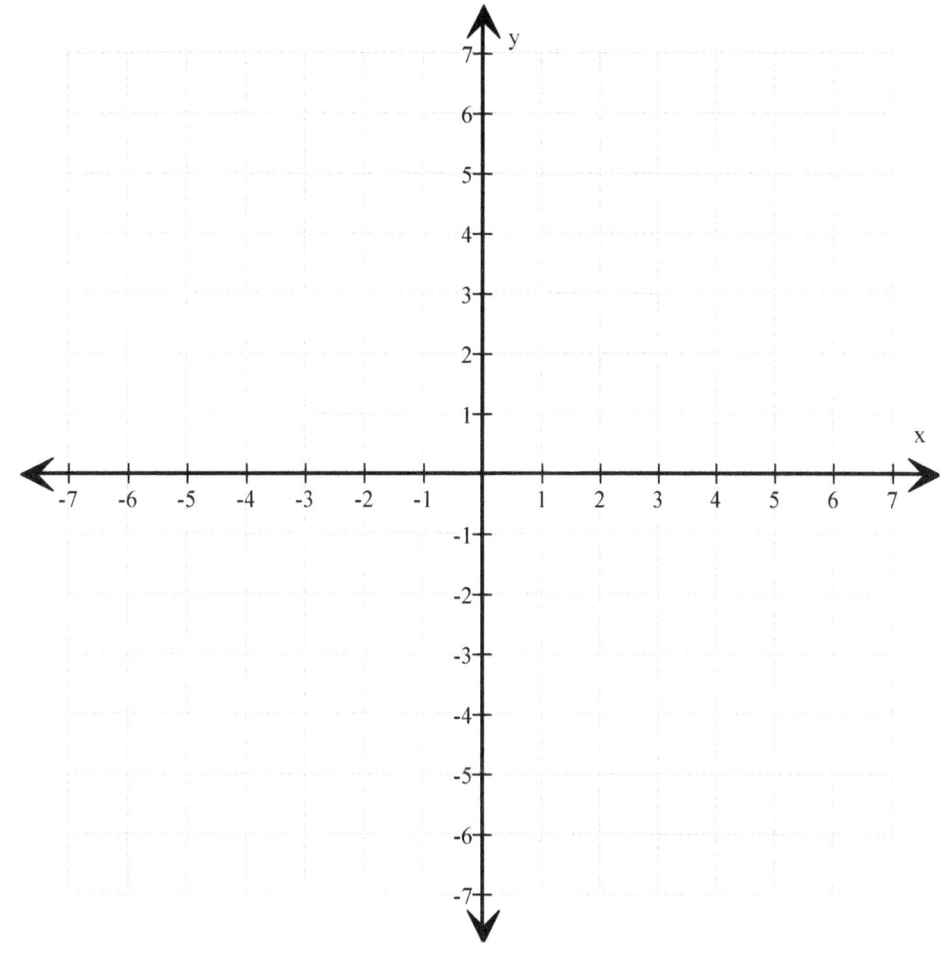

73) PRACTICE: Which coordinate is each person?

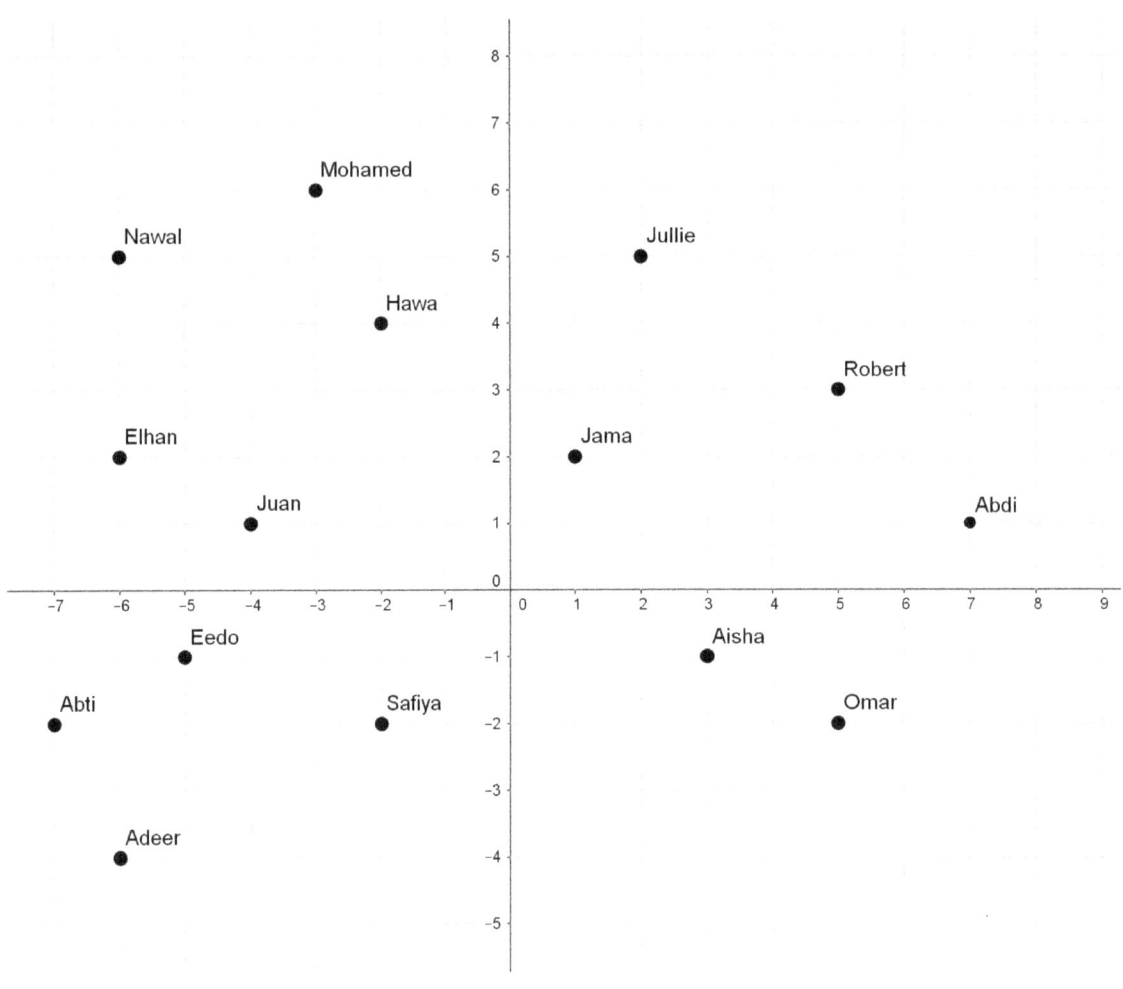

Abti (-7, -2)	Hawa( )	Juan ( )	Omar ( )
Abdi ( )	Eedo ( )	Jullie ( )	Robert ( )
Adeer ( )	Elhan( )	Mohamed ( )	Safiya ( )
Aisha ( )	Jama( )	Nawal ( )	

74) Graphing Linear Equations: How to Find Ordered Pairs

To find the ordered pairs of equation, give any value you like to x, and then solve the y. (Sorry, you can't choose both x and y at the same time).

Example : 2x +y = 8		
If we choose x=0	it means 2 (0) + y = 8	y = 8
If we choose x=1	it means 2(1) + y = 8	y = 6

1) 2x +y = 8

x	y
0	
1	
2	
3	

2) 3x - y = 4

x	y
0	-4
1	-1
2	2
3	5

3) x - y = -2

x	y
0	
1	
2	
-1	

4) x + 2y = 5

x	y
0	2.5
1	2
3	1
5	0

5) y = x - 3

x	y
0	
1	
4	
5	

6) y = 3x - 2

x	y
0	
1	
2	
-1	

7) y = x + 2

x	y
0	
1	
2	
3	

8) y = -4x - 3

x	y
0	
-1	
-2	
1	

75) Graphing Linear Equations: Find Ordered Pairs

Hint: Choose any number for x but choose an easy number for x to find the y number.

1) x + y = 1

x	y

2) x - 2y = 4

x	y

3) x - y = 3

x	y

4) 3x + y = -2

x	y

5) y = 2x - 8

x	y

6) y = x - 4

x	y

7) y = x + 1

x	y

8) y = x

x	y

9) y = x - 3

x	y

10) 2x - y = -2

x	y

11) y = -2x + 4

x	y

12) x + 5y = -10

x	y

13) 3x + y = 6

x	y

14) x - 3y = -3

x	y

15) 2x + y = -8

x	y

16) 2x + y = 0

x	y

76) Find Ordered Pairs for Equations with Fractions

Example 1	$y = \dfrac{1}{2}x + 2$

x	0	2	4	-2	-4
y	2	3	4	1	0

See! It is easier to choose multiples of the 2 in the denominator for the x such as 2,4, 6, -2, -4…

Example 2	$y = \dfrac{2}{3}x - 4$

x	0	3	6	9	-3
y	-4	-2	0	2	-6

See again. It is easier to choose multiples of the 2 in the denominator for the x such as 3,6, 9, -3…

1) $y = \dfrac{2}{3}x + 1$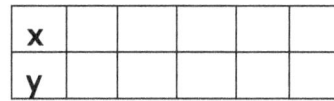

2) $y = \dfrac{2}{5}x - 3$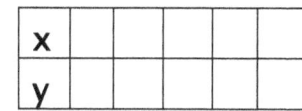

3) $y = \dfrac{1}{4}x + 1$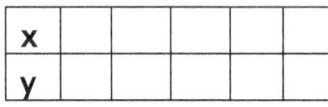

4) $y = \dfrac{3}{2}x - 2$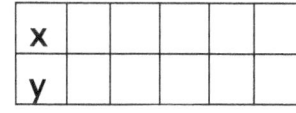

5) $y = \dfrac{1}{3}x + 1$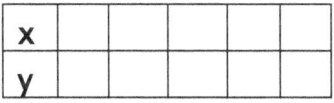

6) $y = \dfrac{1}{5}x - 2$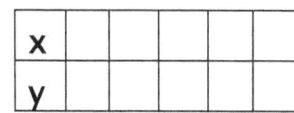

7) $y = \dfrac{3}{4}x + 1$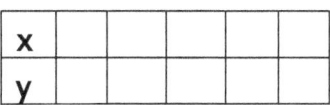

8) $y = \dfrac{4}{3}x - 2$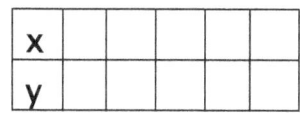

9) $y = \dfrac{1}{6}x - 5$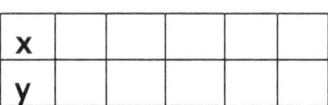

10) $y = \dfrac{2}{7}x + 1$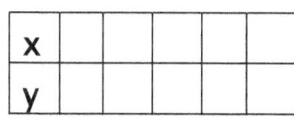

Prepare 2-4 ordered pairs, 2) **Plot** the point 3) **Graph** the line.

Example 1: **Graph y = 2x+3**

Let's find few ordered pairs
and plot on the graph:

x	y
0	3
-1	1
1	5

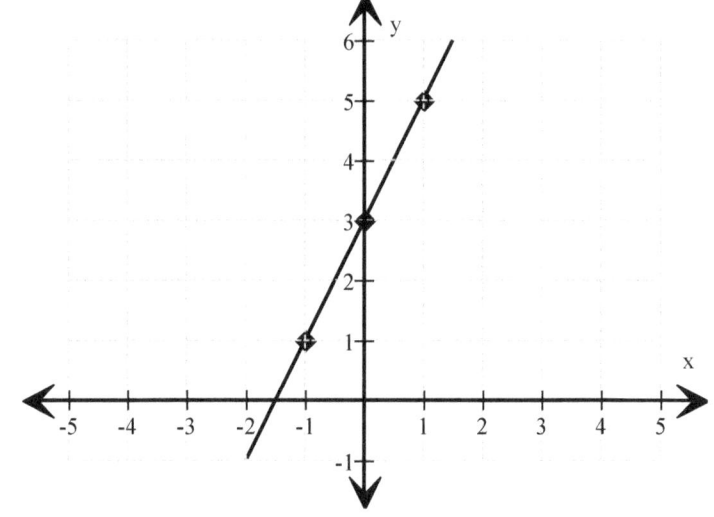

Example 2: **Graph y = -2x+2**

x	Y
0	2
-1	4
1	0

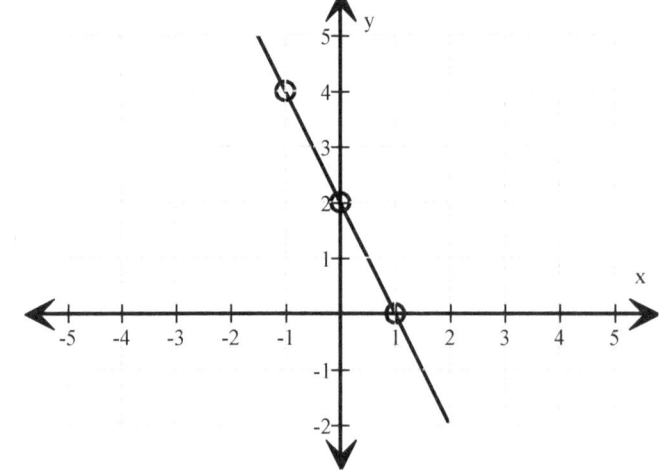

78) Match the Graphs with their correspondent equations

A) y = 2x-3   B) y =-2x - 2   C) y = 3x+1   D) y = x   E) y =-x+3   F) y= -4x+3

1)
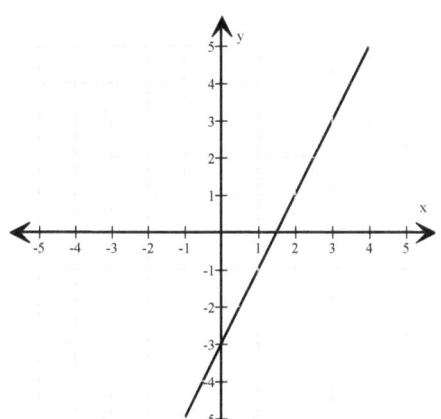

2)
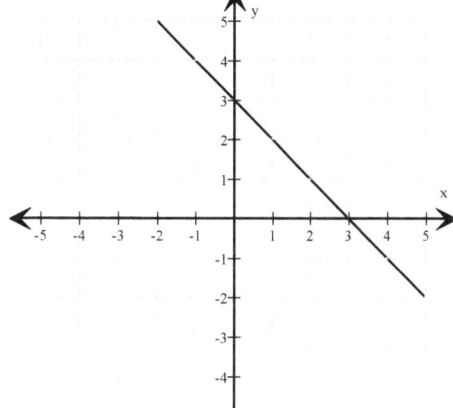

3)
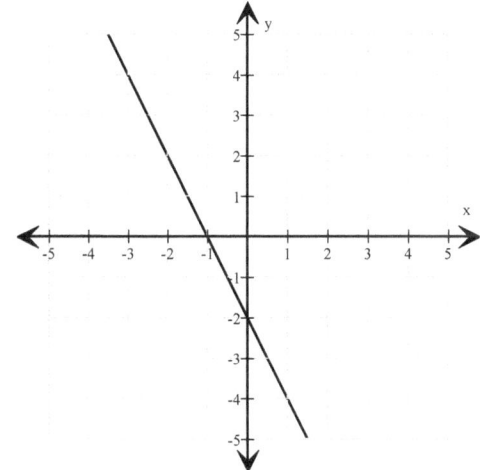

4)
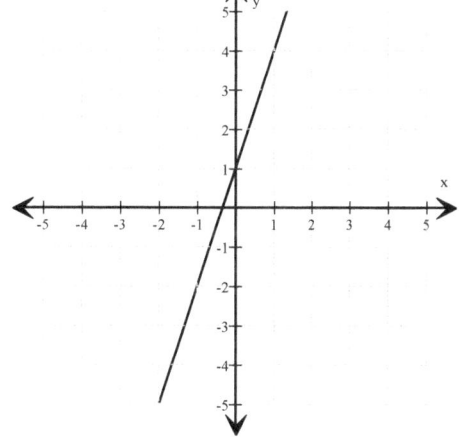

5)
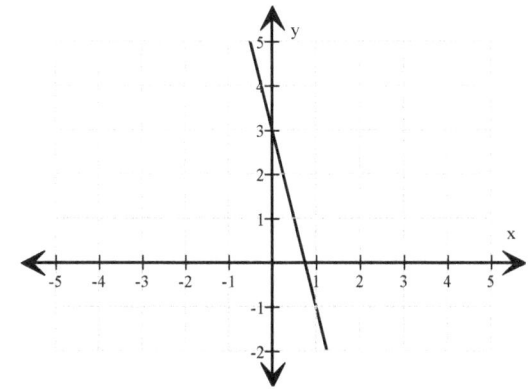

6)
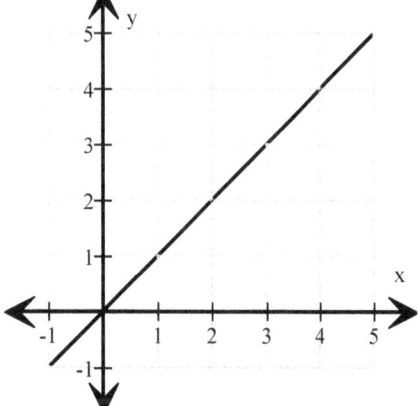

89

79) Match the Graphs with their correspondent equations-2

A) y = 2x-1	**B)** $y = 2x$	C) y =- x	**D)** y =--4x+3	**E)** $y = \frac{1}{3}x + 1$

7)

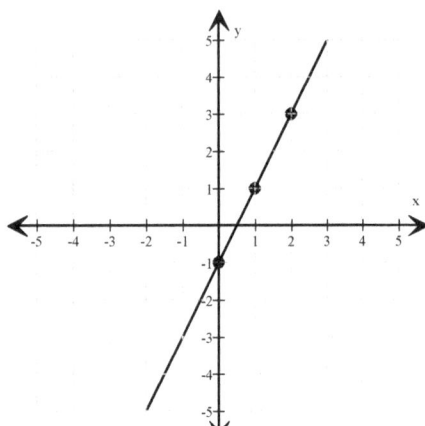

8)

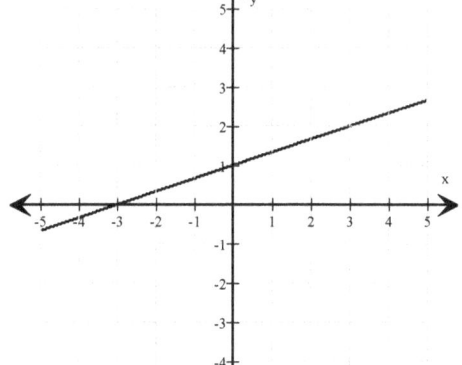

9)

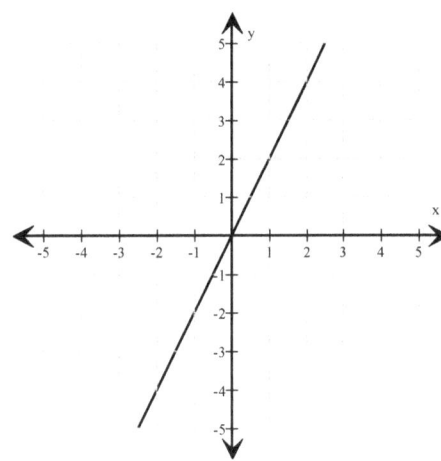

10)

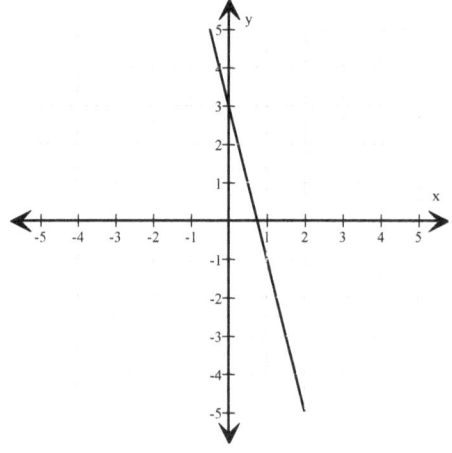

11)

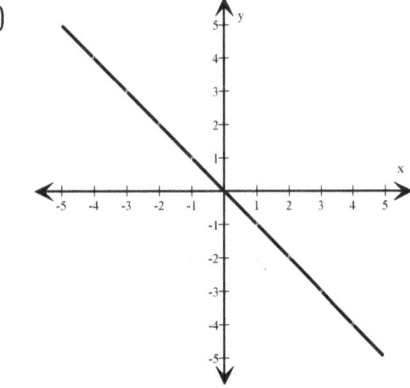

90

80) Graph These Linear Equations

1. Prepare 3 ordered pairs, 2) **Plot** the points 3) **Graph** the line:

1) $y = 2x$

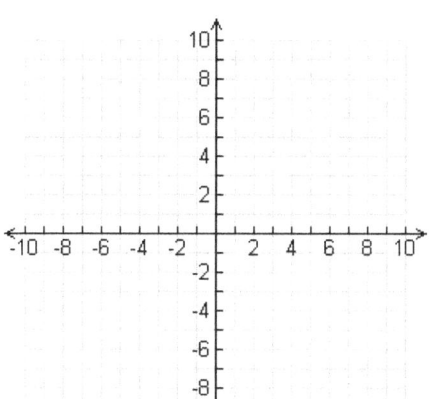

2) $y = x - 2$

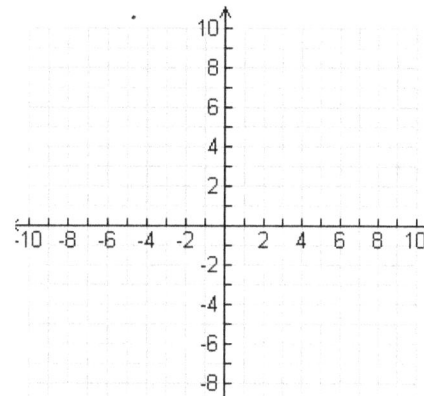

3) $y = -4x - 2$

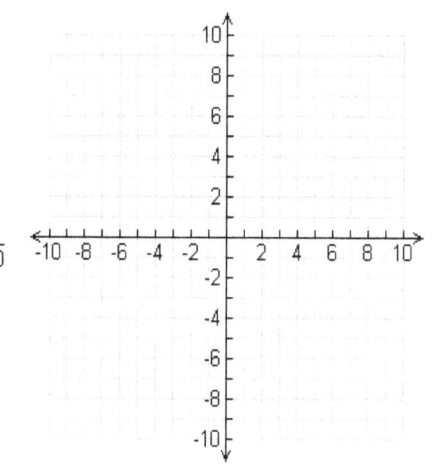

4) $y = x + 2$

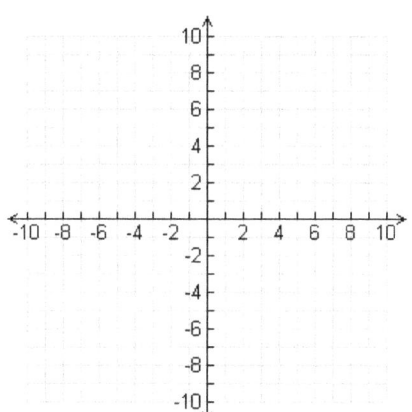

5) $y = 3x + 1$

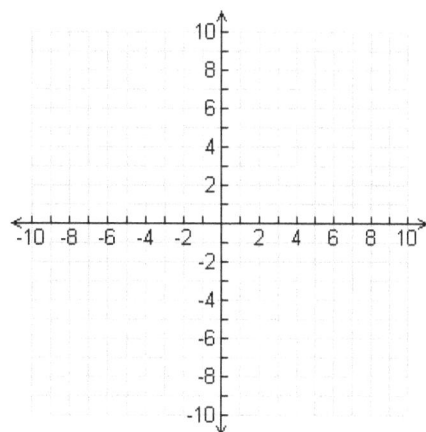

6) $2x + 3y = 6$

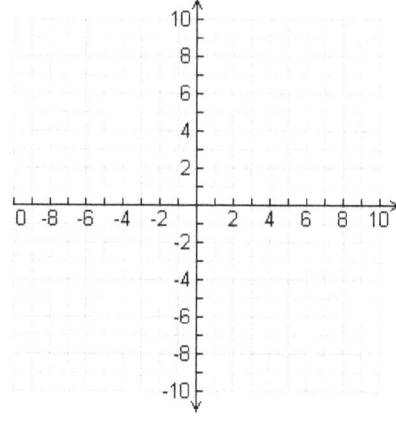

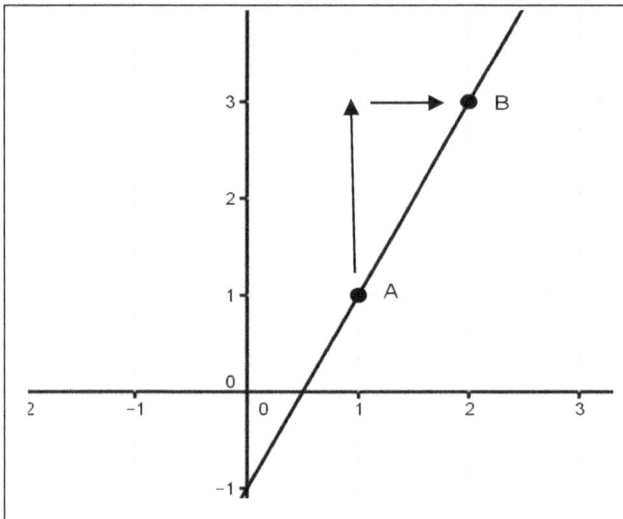

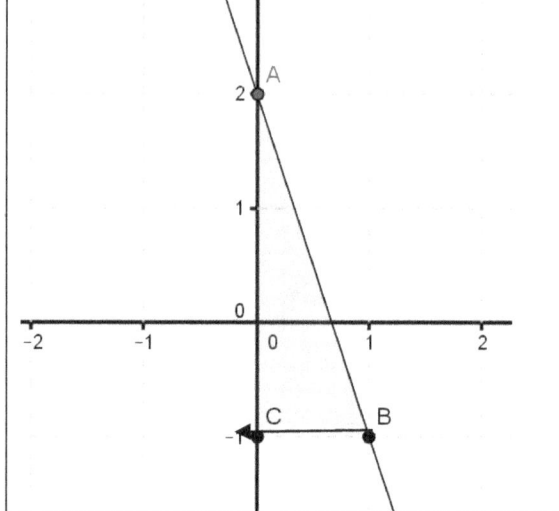

To get from point A to B rise up (at y axis) by 2 points and run (on x axis) by 1:

$$Slope = \frac{Rise}{Run} = \frac{2}{1} = 2$$

Rise from point C and A. = 3
Run from B to C= -1 (It is negative to move backwards)

$$Slope = \frac{Rise}{Run} = \frac{3}{-1} = -3$$

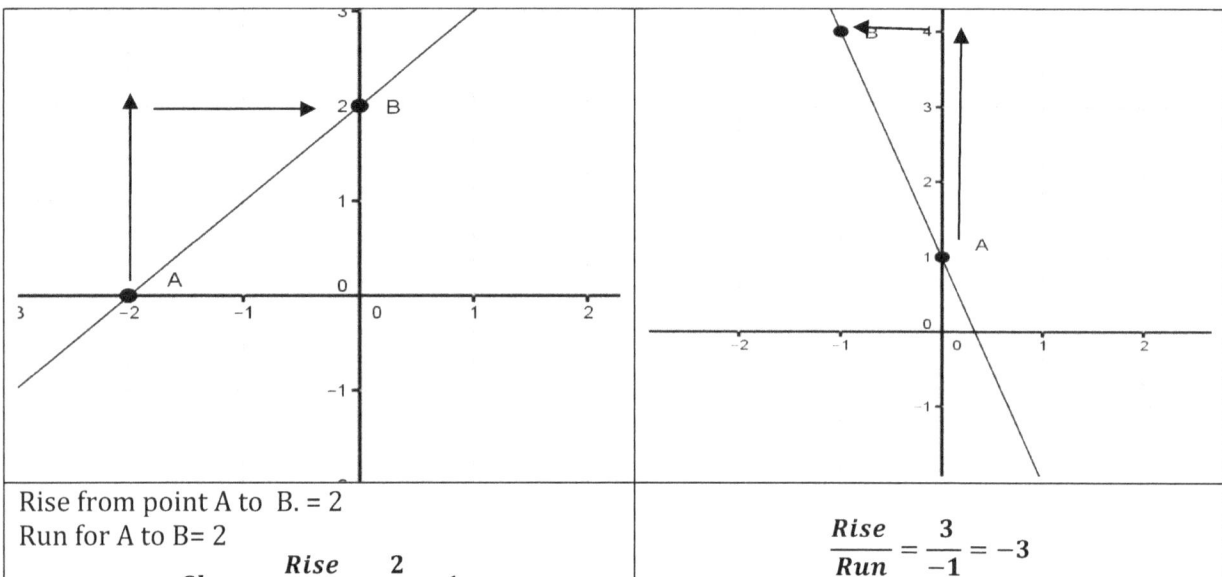

Rise from point A to B. = 2
Run for A to B= 2

$$Slope = \frac{Rise}{Run} = \frac{2}{2} = 1$$

$$\frac{Rise}{Run} = \frac{3}{-1} = -3$$

82) Find the Slope from the Graphs

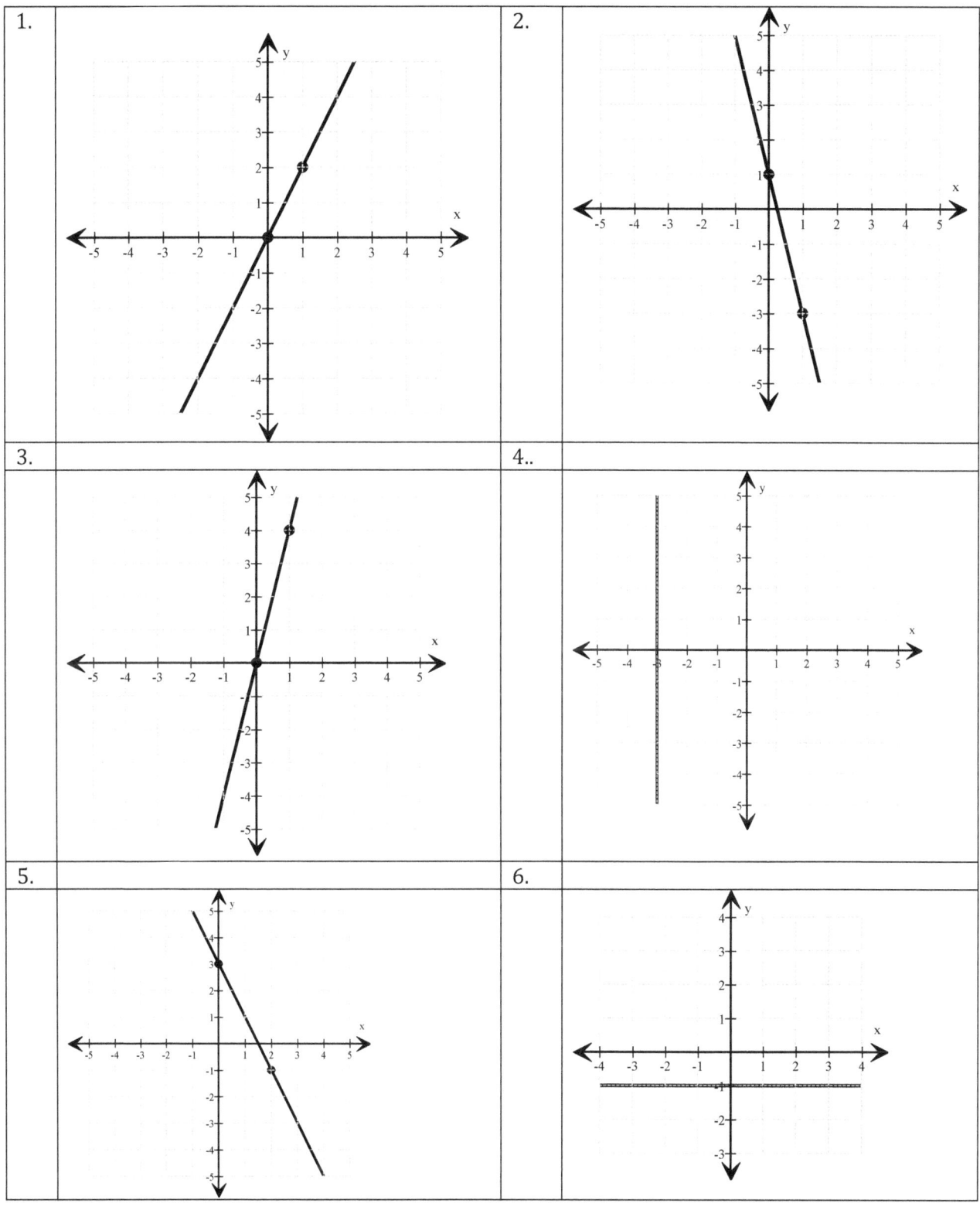

93

1)

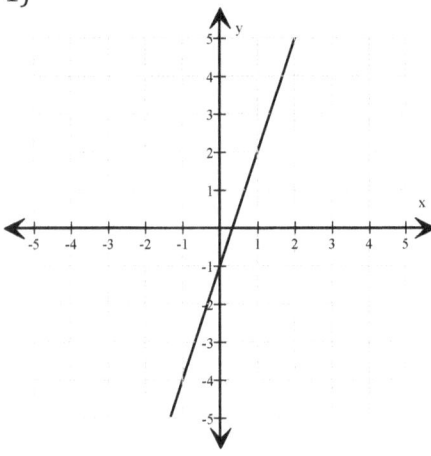

2)

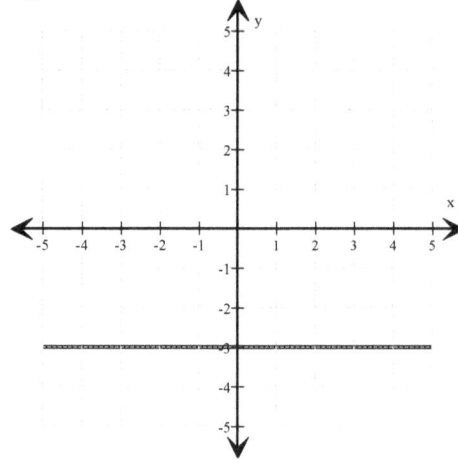

3)

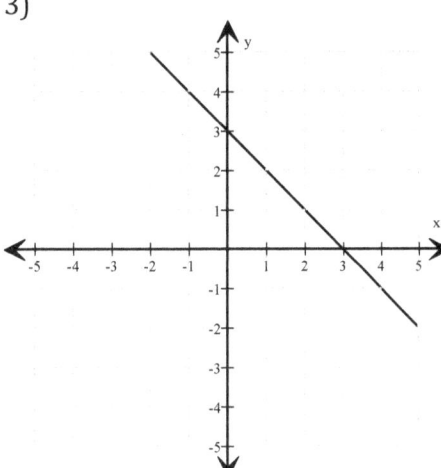

4)

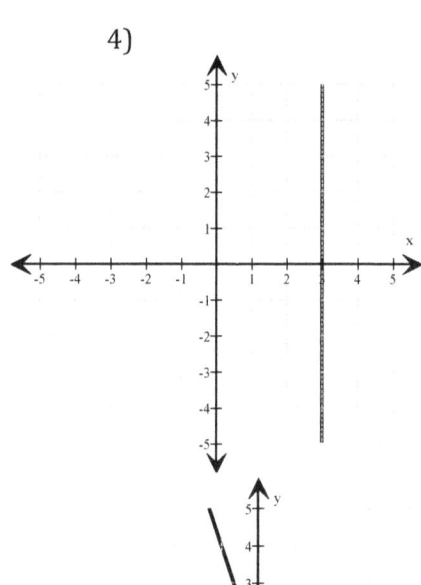

5)

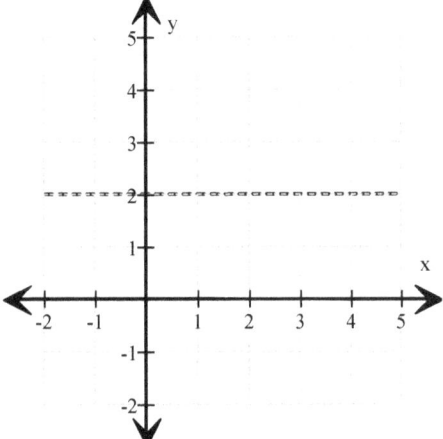

6)

84) Find the Slope from Ordered Pairs

Find the slope that passes through each pair of ordered points: See examples.

	Example1 : (O,3),(4,1)	Example 2: (2,-3), (-4,6)
1) Label it:	$x_{1=0}, x_{2=4}$ $y_{1=3}, y_{2=1}$	$x_{1=2}, x_{2=-4}$ $y_{1=-3}, y_{2=6}$
2) Put it in the Formula: $m = \dfrac{y_2 - y_1}{x_2 - x_1}$	$m = \dfrac{1-3}{4-0} = \dfrac{-2}{4} = -\dfrac{1}{2}$	$m = \dfrac{6-(-3)}{-4-2} = \dfrac{9}{-6} = -\dfrac{3}{2}$
	Remember: m stands for slope. Also, watch out for the signs!	

1. (2, 3), (4,1) _____

2. (-4, 3), (5,1) _____

3. (1, 3), (2,4) _____

4. (5, 5), (6, 7) _____

5. (3, 3), (-4, 1) _____

6. (2,-3), (4, 7) _____

7. (3,-3), (-4, 6) _____

8. (7,-3), (-4, 8) _____

9. (1, 4), (4, 7) _____

10. (0,-3), (-4, 5) _____

11) (2,-0), (-1, 2) _____

12) (2,7) , (7, 6) _____

13) (7,-3), (6, 6) _____

14) (7,-3), (0, 6) _____

15) (2,-3), (-4, 6) _____

16) (5,-3), (3, 6) _____

17) (2,-3), (1, 2) _____

18) (4,-3), (3, 8) _____

19) (2,-3), (-4, 9) _____

20) (2, 9), (-5, 6) _____

85) Write the Equation of the Line: Use Slope/Y-intercept:

A) Use the given slope and the y-intercept to **write the equation of the line** in Slope Intercept form $(y = mx + b)$. See examples:

Example1 :	**Example 2 :**
Slope =3 y-intercept =1	**Slope =-1 y-intercept =0**
$y = mx + b$ $y = 3x + 1$	$y = mx + b$ $y = -x + 0$ Or simply: $y = -x$

Remember: "m" is the slope and "b" stands for the y-intercept.

1) Slope =2, y-intercept = 1

2) Slope =-2, y-intercept = 2

3) Slope =-4, y-Intercept = -2

4) Slope =-1, y-intercept = 7

5) Slope = 5, y-Intercept = 0

6) Slope =-6, y-intercept = 11

7) Slope = 0, y-Intercept = -5

8) Slope =$\frac{2}{3}$, y-intercept = -38

9) Slope = -1, y-Intercept = -3

10) Slope =-10 , y-Intercept = -27

B) Given the equation of the line, find the slope and the Y-intercept. See Example:

11) $y = 3x + 1$	$m = 3 ;\ b = 1$	12) $y = 7$	
13) $y = -5x + 7$		14) $y = 8x$	
15) $y = -6x - 4$		16) $y = -3x - 1$	
17) $y = x$		18) $y = 2x + 3$	
19) $y = -9x + 8$		20) $y = -4x - 5$	

86) Graph the Line from slope and y-intercept

Example: Graph the line with the given equation;
$$y = 2x - 1$$
First: see the slope $= 2$ and the y-intercept $= -1$
So, mark the y-intercept (or -1).

Next: go up 2 points and over 1 point to the right.
1) Connect the point (dots)!

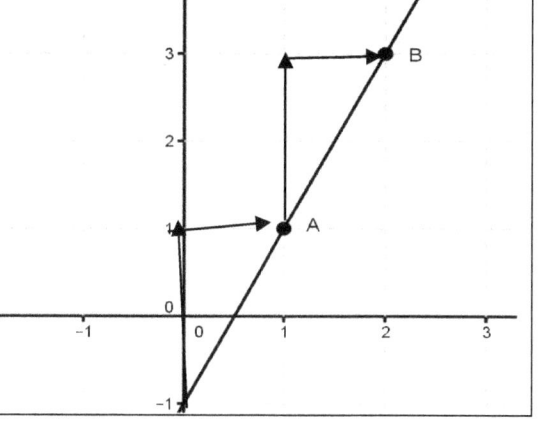

$y = 4x - 2$

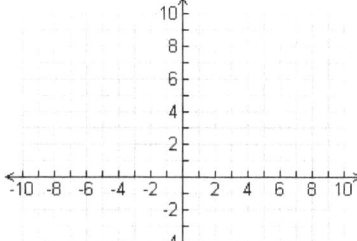

$y = x + 1$

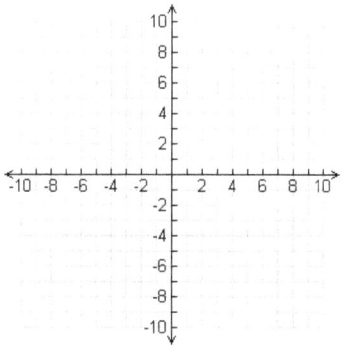

$y = 2x + 2$

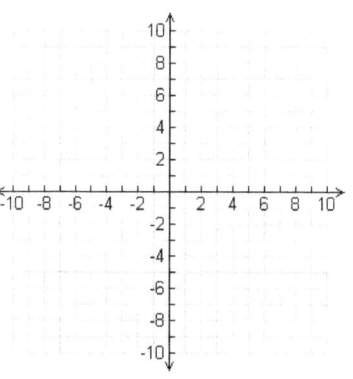

$y = -2x - 1$

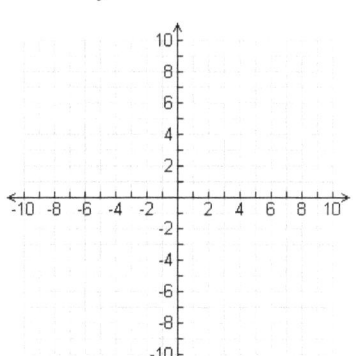

$y = 3x$

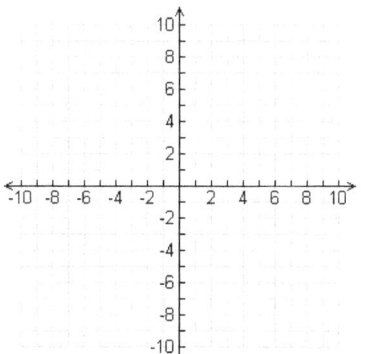

$y = 6x - 6$

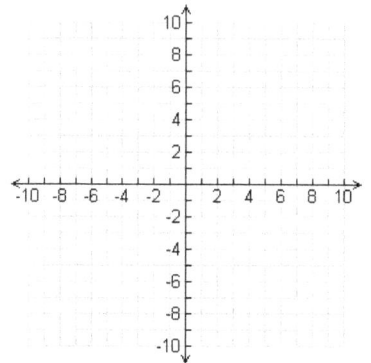

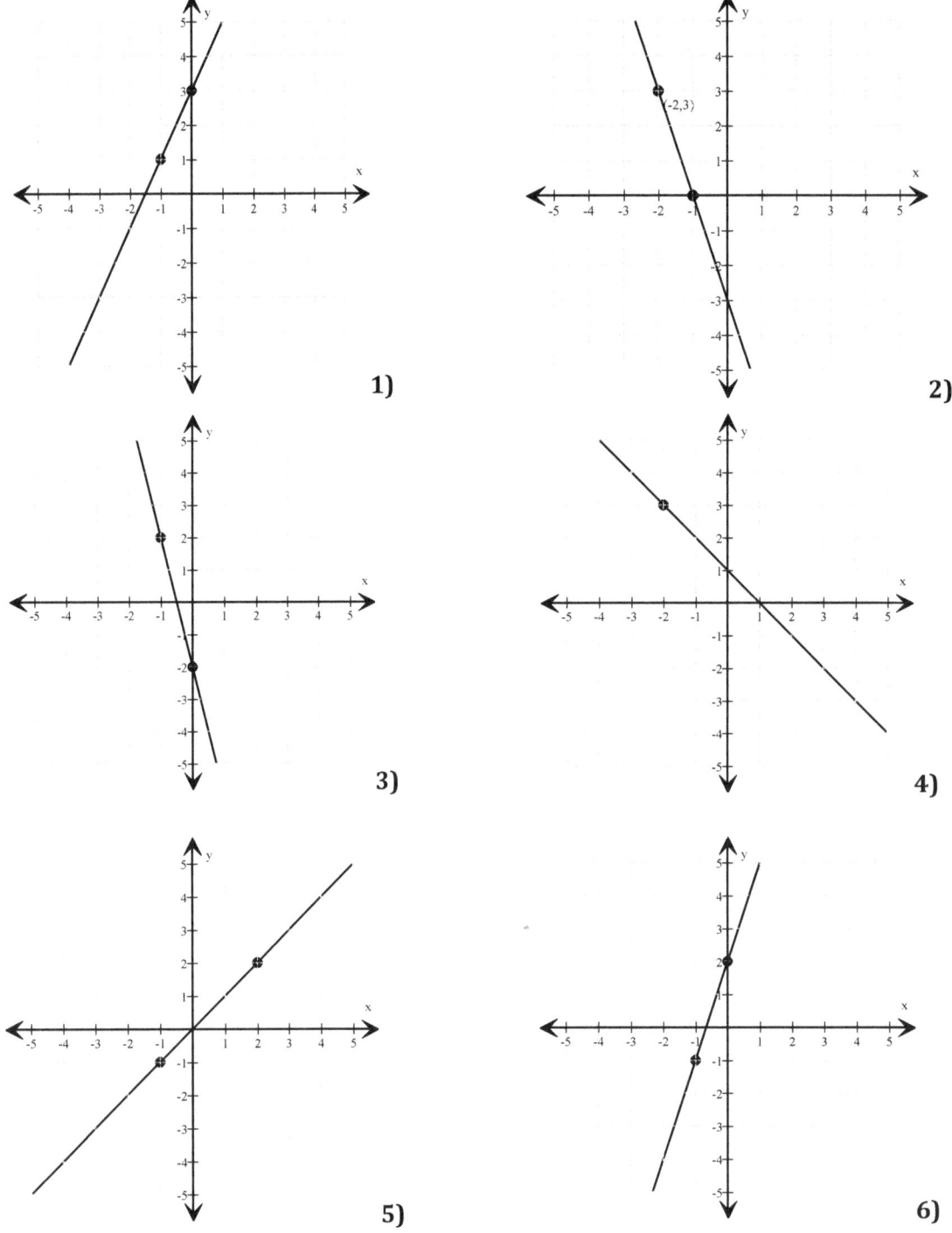

1)

2)

3)

4)

5)

6)

88) The Pythagorean Theorem

The longest side of a triangle is called the **hypotenuse**. The two other sides are called **legs.** In each right angle triangle, the squares of the legs is equal to the square of hypotenuse.

$$hypotenuse^2 = leg1^2 + leg2^2$$

Mostly hypotenuse is shortened for c and the legs as a, and b:

$$c^2 = a^2 + b^2$$

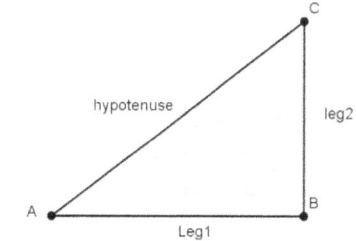

Example 1: Given a= 3, b=4 find the hypotenuse:

$$c^2 = a^2 + b^2$$
$$c^2 = 3^2 + 4^2 = 9 + 16 = 25$$
$$c = \sqrt{25} = 5$$

Example 2: Given c= 5, b=4 find the other leg of the triangle:

$$5^2 = a^2 + 4^2$$
$$5^2 - 4^2 = a^2$$
$$a^2 = 25 - 16 = 9$$
$$a = \sqrt{9} = 3$$

If c is the hypotenuse, find the missing sides. Use two decimal places as needed.

1) $a = 8$ b = 6   c =?

2) a =?   b = 30   c = 40

3) a =?   b = 6   c = 9

4) a =?   b = 5   c = 8

5) $a = 15$ b?   c = 20

6) $a = 36$ b =?   c = 49

7) $a = 7$ b = 4   c =?

8) $a = 8$ b = 6,   c =?

89) The Distance Formula

Example: What is the length of segment (a) or the distance between points A and B?

The distance between any two points is calculated by using the distance formula

$$distance = \sqrt{(x_2 - x_1)^2 + (y_2 - y_1)^2}$$

$$d = \sqrt{(5 - 1)^2 + (4 - 1)^2}$$

$$d = \sqrt{16 + 9}$$

$$d = 5$$

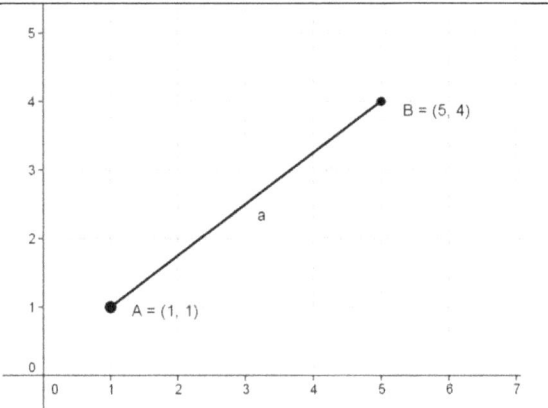

Calculate the distance of the following pairs of points. Use a calculator if you need to!

1. A (1,2), B(3,4)

2. C (4,5), D (6,4)

3. E (7,2), F(4,6)

4) G (-2,-3), H(7,-6)

5)I (-4,5), J (-8,8)

6) K(-3,-3), L(6,6)

7) M (-5, -6) N (2,4)

8) P (-6,4), Q (4,8)

9) R (5,10) S (-5,-6)

90) POLYGONS IN SNAP SHOT!

Definition: A polygon means a closed plane figure with three sides or more. So, we can generally divide it into three groups: triangles (three sided) Quadrilaterals (four Sides), and Figures with 5 sides or more!

1) Triangles: Are three Sides Figures. They could include Equilateral, Isosceles, right and Scalene

EQUILATERAL- All Sides are equal

ISOSCELES- 2 sides are equal

RIGHT ANGLED: One angle is 90 degrees

SCALENE- All sides are Different

2) Quadrilaterals: Are Four sides figures. They Include: (a) **parallelogram** (rectangle, rhombus, square), (b) Trapezoid, and (c) **Kite**

PARALLELOGRAMS: Each opposite sides are equal. Opposite angles are also equal

TRAPEZOID: has only one pair of parallel sides

KITE: Has 2 pairs of adjacent sides that are equal.

3) Five Sided and More: Examples of common figures of these are

PENTAGON (5 sides)   *HEXAGON* (6 sides),

HEPTAGON (7 sides)   *OCTAGON* (8 sides),

NONAGON (9 sides)   *DECAGON* (10 sides)

Polygons In Three Groups

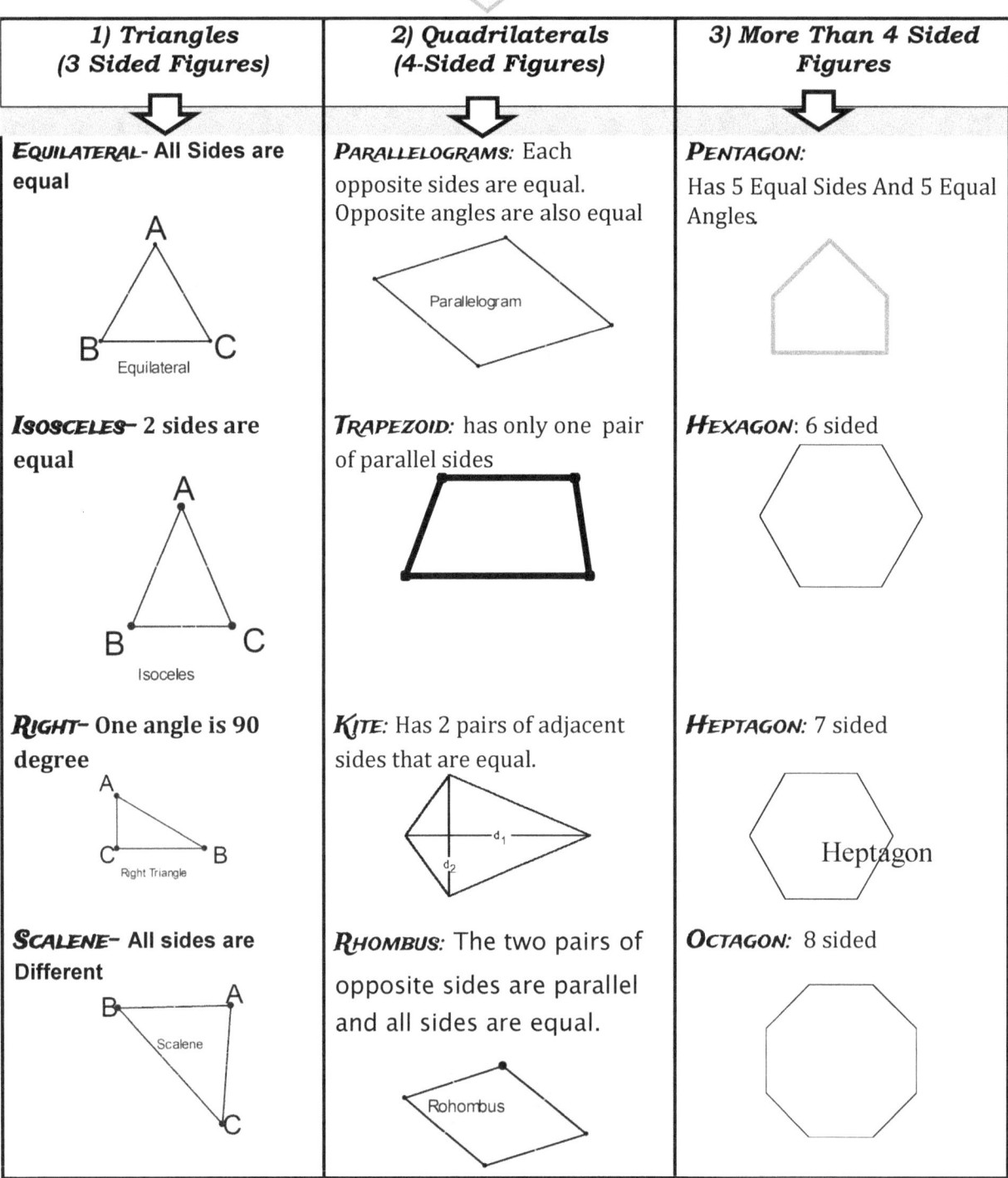

1) Triangles (3 Sided Figures)	2) Quadrilaterals (4-Sided Figures)	3) More Than 4 Sided Figures
EQUILATERAL- All Sides are equal	**PARALLELOGRAMS**: Each opposite sides are equal. Opposite angles are also equal	**PENTAGON**: Has 5 Equal Sides And 5 Equal Angles.
ISOSCELES- 2 sides are equal	**TRAPEZOID**: has only one pair of parallel sides	**HEXAGON**: 6 sided
RIGHT- One angle is 90 degree	**KITE**: Has 2 pairs of adjacent sides that are equal.	**HEPTAGON**: 7 sided
SCALENE- All sides are Different	**RHOMBUS**: The two pairs of opposite sides are parallel and all sides are equal.	**OCTAGON**: 8 sided

A **perimeter (P)** is the distance around the polygon. To find it, add the lengths of all sides. **The Area (A)** of Polygons can be found in different ways (See Examples below):

Examples: Find the area and the perimeter for each of the following polygon (Assume all units are in feet).

Area of square $= side^2$
$$A = 5ft \times 5ft = 25ft^2$$

Perimeter (P) $= 5 \times 4 = 20ft$
(Since there are four equal sides)

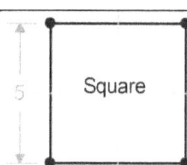

Area of a triangle $= \dfrac{\text{height} \times \text{Base}}{2}$

$$A = \frac{3 \times 2}{2} = 3ft^2$$
$$P = 3 + 4 + 2 = 9\,ft$$

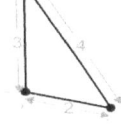

Area of Rectangle = Length × Width
$$A = L \times W$$

$$= 7 \times 4 = 28ft^2$$
$$P = 2L + 2W = 2 \times 4 + 2 \times 7$$
$$P = 8 + 14 = 22ft$$

Area of a trapwezoid

$$= \frac{1}{2}\text{height}\,(\text{Base1} + \text{Base2})$$
$$A = \frac{1}{2}h\,(b_1 + b_2)$$
$$A = \frac{1}{2} \times 4\,(3 + 6) = 18\ ft^2$$
$$P = 3 + 5 + 4 + 6 = 18ft$$

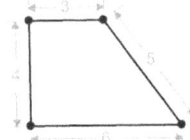

Kite:
   A Kite has **two diagonals** $d_1 \text{ and } d_2$ that are perpendicular with each other.
   The area of a kite is the total space inside the boundary of a kite.

Area of the kite $= \dfrac{1}{2}d_1 \times d_2 = \dfrac{1}{2}\,(10) \times (4) = 20$

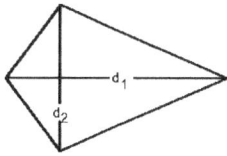

Assume in this example d1= 10 and d2 = 4

Area and Perimeter Practice

Problems 1 to 9: Find the areas and the perimeters. Unless shown, assume all units are inches:

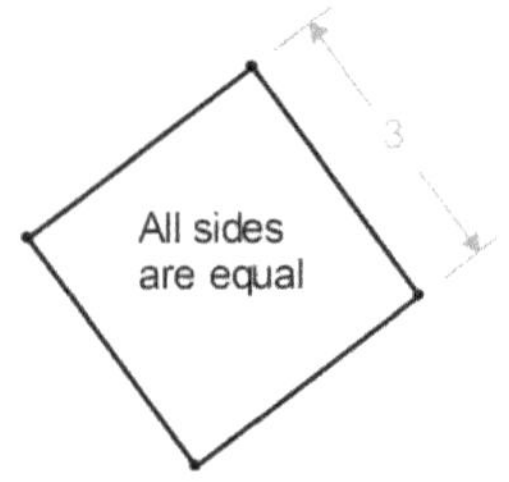

1)
All sides are equal
3

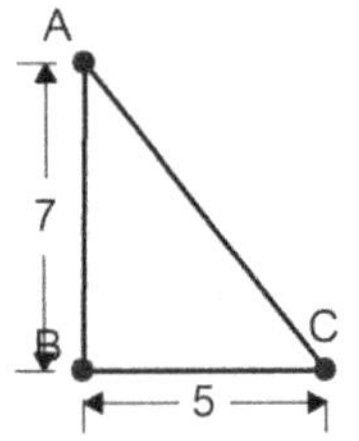

2)
A
7 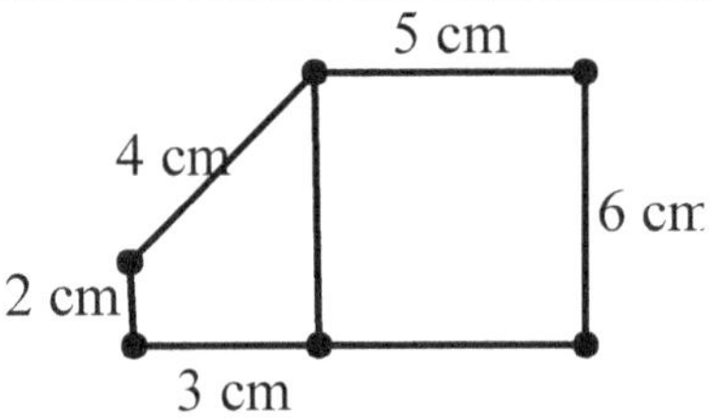
B
5
C

3)
10
5
4
7

4)
A
8
D
5
B
5
C

5)
8 in
8 in
14 in

6)
6 cm
2 cm
3 cm
6 cm
8 cm

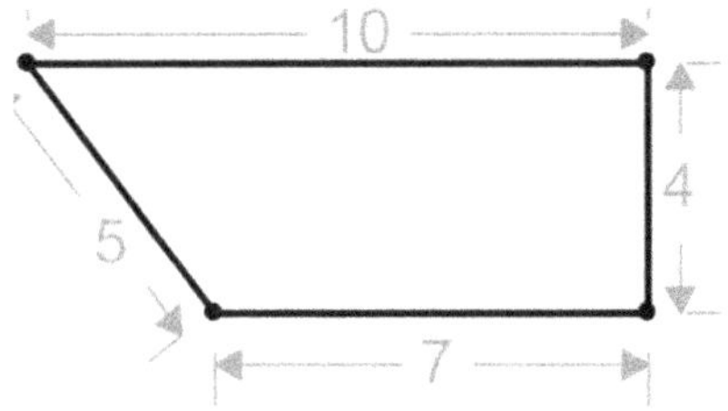

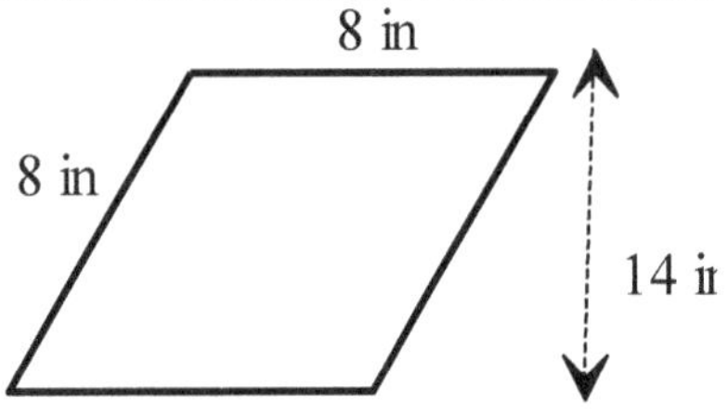

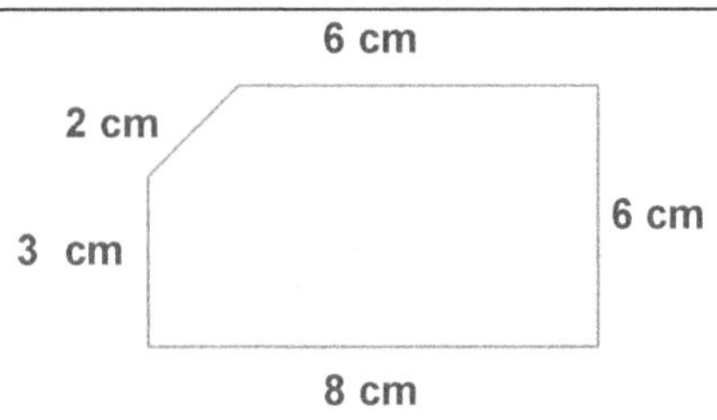

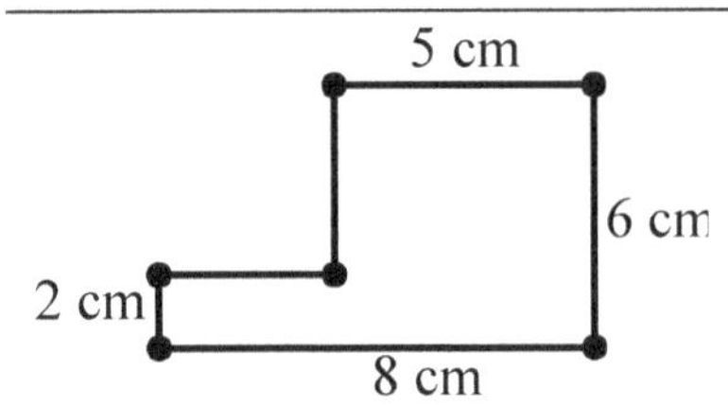

7)
5 cm
6 cm
2 cm
8 cm

8)
5 cm
4 cm
6 cm
2 cm
3 cm

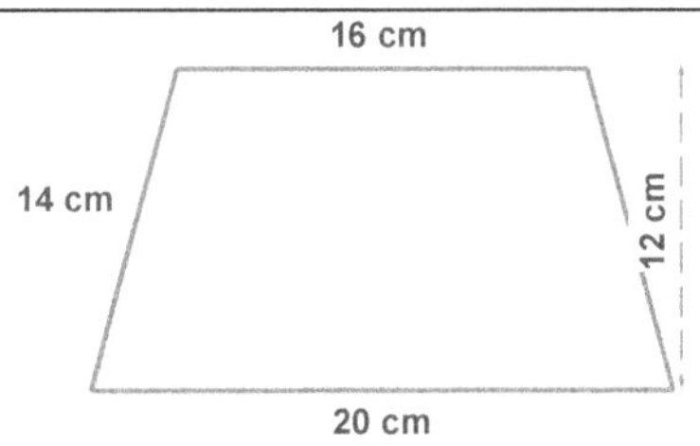

9)
16 cm
14 cm
12 cm
20 cm

The premeter is 18, find the length of the missing side HK?

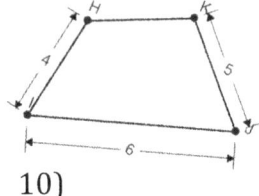

10)

(a) Find the area of triangle JHG
(b) Find the area of the trapezoid.

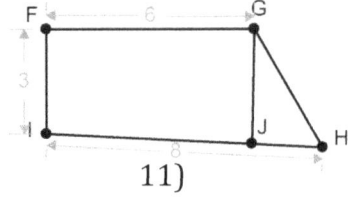

11)

104

93) Circles: AREAS and Circumference

A) What is a circle? How it is named?

A circle is a shape that has all of its points the same distance from the center. Circles are named after their centers. So, for example we have here Circle A and Circle C. (Below).

B) The Diameter and The Radius

The Diameter (d) connects side to side of the circle passing through the center.
The Radius (r) is half of the diameter or the distance from the center to any points in the circle.

$$r = \frac{d}{2}$$

C) The Circumference and the Area:

The circumference (C) is like the perimeter: It is the distance around the circle.

$$C = 2\pi r \text{ or } C = \pi d$$

r is the radius

$\pi = 3.14$

The Area (A) of the circle is the area enclosed by the circle.

$$A = \pi r^2$$

D) Calculations: Example:

Calculate the diameter, the circumference and the area of a circle with a radius of 6 cm.

$$d = 2r = 2 \times 6 = 12 \text{ cm}$$
$$C = \pi d = 3.14 \times 12 cm = 37.68 cm$$
$$A = \pi r^2 = 3.14 \times (6cm)^2 = 18.84 cm^2$$

93) Circle Practice

Find the circumference, and the area of the following circles:

1) $d = 4cm$ $\qquad C = ?$ $\qquad A = ?$

2) $r = 5\ cm$ $\qquad C = ?$ $\qquad A = ?$

3) $d = 14\ cm$ $\qquad C = ?$ $\qquad A = ?$

4) $r = 7cm$ $\qquad C = ?$ $\qquad A = ?$

5) Name the Circles and then find the length of the diameter, the circumference and the area of each circle.

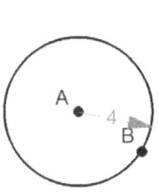 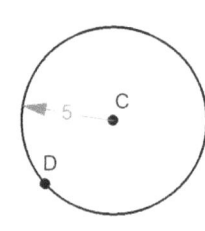 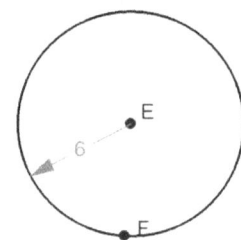

6) **True or False**: A circumference of a circle can be found by multiplying 3.14 with the diameter of the circle.

7) **True or False**: A circumference of a circle is π times its radius?

8) A pizza has a diameter of 8 inches. Find its (a) circumference and

    b) Its area.

1) Find the area of a kite if $d_1 = 10\ cm\ and\ d_2 = 4\ cm$

A kite has **two diagonals** $d_1 and\ d_2$ that are perpendicular with each other. They bisect at right angles.

The area of a kite is the total space inside the boundary of a kite.

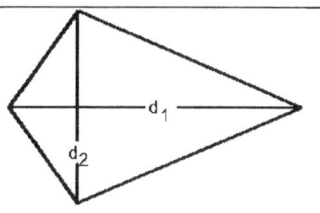

$$\textbf{Area of the kite}\ = \frac{1}{2}d_1 \times d_2$$
$$\textbf{Area of the kite}\ = \frac{1}{2}\ (10) \times (4) = 20$$

Find $d_2\ if\ d_1 = 12\ cm\ and\ the\ area\ is\ 60cm$

$$\textbf{Missing diagonal}\ = \frac{2\ \times Area}{known\ diagonal} = \frac{2 \times 60}{12} = 10cm$$

Find the surface area of a kite with the given diagonals.
1) $d_1 = 40\ cm\ and\ d_2 = 10\ cm$

2) $d_1 = 100cm\ and\ d_2 = 20cm$

3) $d_1 = 18\ cm\ and\ d_2 = 14cm$

Find the length of the missing diagonal:.

4) $d_1 = 4\ cm\ and\ Area = 10\ cm$

5) $d_2 = 5\ cm\ and\ Area = 4\ cm$

6) $d_1 = 20\ cm\ and\ Area = 80\ cm$

95) Finding Surface Area and Volume of a Cube

Example: Find (a) the surface area and (b) the volume of the cube that is 8 cm in each side

$a)$ *Surface Area of a cube with side S* $= 6S^2$

   Surface area $= 6 \times 8^2 = 6 \times 64 = 384 cm^3$

$b)$ *Volume of a cube with side S* $= S^3$

   $Volume = 8^3 = 512 cm^3$

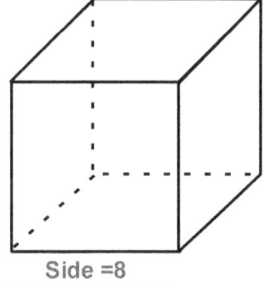

Side =8

Find the surface area of each of the following cubes with the given side. Show your calculations

1) S =5   2) S= 8   3) S= 10

4) S = 6   5) S= 12   6) S= 13

Find the volume of each of the following cubes.

7) S = 10   8) S= 4   9) S= 9

96) Finding Surface Area and Volume of Rectangular Prism

Example : Find (a) the surface area and (b) the volume of a rectangular prism with the height of 3ft., length of 6 ft., and width of 4 ft.

a) $Surface\ Area = 2(h \times l + h \times w + l \times w)$
$surface\ Area = 2(3 \times 6 + 3 \times 4 + 6 \times 4)$
$\qquad\qquad = 2(18 + 12 + 24)$
$\qquad\qquad = 2(54) = 108\ in^2$

b) $Volume = l \times w \times h$
$\qquad Volume = 6ft \times 4ft \times 3\text{ft}.$
$\qquad\qquad = 72ft^3$

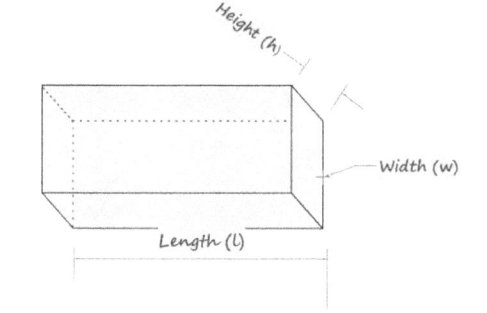

Find the surface area of a rectangular prism with the following dimensions. Make sure to round it to the nearest whole number.

1) Height = 6ft; length= 4ft; width = 5 ft.

2) Height = 4ft; length= 7ft; width = 3 ft.

3) Height =10ft; length =3ft; width = 2 ft.

4) Height = 12ft; length= 5ft; width = 5 ft.

Find the volume of a rectangular prism with the following dimensions:

5) Height = 6 cm; length= 4 cm; width = 5cm

6) Height = 12 cm; length= 10 cm; width = 8 cm

7) Height = 9 cm; length= 3 cm; width = 12cm

8) Height = 11 cm; length= 9cm; width = 10 cm

97) Finding Surface Area of Sphere

a) **Surface Area** $= 4\pi r^2$
$$= 4 \times 3.14 \times 5^2 = 314 cm^2$$

b) **Volume** $= \dfrac{4}{3}\pi r^3 =$
$$= \dfrac{4}{3} \times 3.14 \times 5^3 = 523.3\ cm^3$$

radius (r)

Find the surface area of each of the following sphere with the given radius in cm. Round your answer to the nearest whole number.

1) r = 3

2) r = 8

3) r = 10

4) r = 7

5) r = 4

6) r = 12

Find the volume of each of the following spheres with the given radius. Show your calculations

7) r = 2

8) r = 7

9) r = 5

Example : Find (a) the surface area and (b) the volume of a **cylinder** with radius of 3 cm and height of 10 cm.

a) **Surface Area** $= 2\pi r^2 + 2\pi rh$
$\qquad\qquad = 2 \times 3.14 \times 3^2 + 2 \times 3.14 \times 3 \times 10$
$\qquad\qquad = 56.52 + 188.4 = 244.92 cm^2$

b) **Volume** $= \pi r^2 h = 3.14 \times 3^2 \times 10 = 282.6\ cm^3$

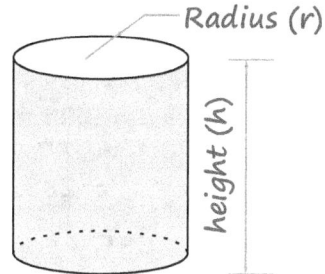

Radius (r)

height (h)

Find the surface area of each of the following cylinder with the given radius and height. The dimensions are all in meters (m). Round your answer to the nearest whole number.

1) r = 3; h = 8

2) r = 4; h = 10

3) r = 10; h = 20

4) r = 7; h = 12

5) r = 11; h = 21

6) r = 12; h = 16

Find the volume of each of the following cylinder with the given radius and height. The dimensions are all in meters (m). Round your answer to the nearest whole number.

7) r = 2 ; h = 8

8) r = 6; h = 30

9) r = 8; h = 12

99) Find Surface Area and Volume of a Cone

Example: Find (a) the surface area and (b) the volume of a cone with slant length of 10 ft. , height of 6 ft. and radius of 5 ft.

a) $Surface\ Area = \pi rs + \pi r^2$
$$= 3.14 \times 5 \times 10 + 3.14 \times 5^2$$
$$= 157 + 78.5$$
$$= 235.5\ cm^2$$

b) $Volume = \frac{1}{3}\pi r^2 h$
$$= \frac{1}{3} \times 3.14 \times 5^2 \times 6$$
$$= 157$$

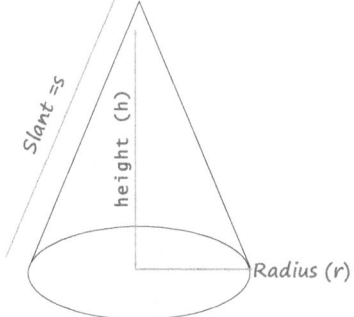

Find the surface area of a cone with the given dimensions. All dimensions are in ft.

1) r = 3; s = 8

2) r = 4; s = 10

3) r = 5; s = 20

4) r = 7; s = 12

5) r = 6; s = 21

6) r = 8 ; s = 16

Find the volume of a cone with:

7) r = 2 ; h= 8;

8) r = 6; h= 30

9) r = 8; h= 2

100) Finding Surface Area and Volume of Pyramid

Find the surface area and volume of the pyramid shown:

Assume base =5 cm ;   Slant =10 cm;   Height= 8 cm

Surface Area =

$$\text{Base area} + \frac{\textbf{perimeter of base} \times \textbf{slant height}}{2}$$

$$base\ area = 5 \times 5 = 25$$

$$Perimeter = 4 \times 5 = 20 \text{ (four sided)}$$

$$Surface\ Area = 25 + \frac{20 \times 10}{2} = 125\ cm^2$$

$b)\ Volume = \dfrac{1}{3} \times base\ area \times height$

$$= \tfrac{1}{3} \times 25 \times 8 = 66.7 cm^3$$

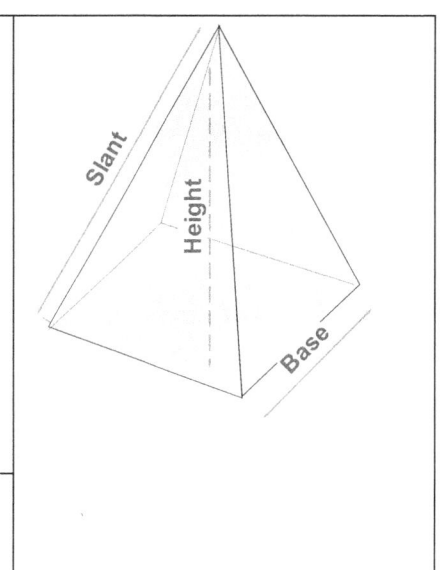

Find the surface area of the square pyramid with the given slant and base. Use 3.14 for π.

1) base = 4 cm;   slant = 10 cm

2) base = 3 cm; slant = 9 cm

3) base = 4 cm; slant = 9 cm

Find the Volume of the square pyramid with the given slant, base and height. Use 3.14 for π.

4) base = 5 cm;   height =8 cm.

5) base = 3 cm;   height =8 cm.

6) base = 4 cm;   slant = 9 cm;   height =8 cm.

101) Review of three Dimensional Area and Volumes

Find the surface area of the following figures with the given dimensions:

1) **Cube**: Side length = 15 cm.

2) **Rectangular prism**: height = 12ft; length = 6ft;   width = 4ft

3) **Sphere**: radius of 11 cm.

4) **Cylinder**: radius of 10 m and a height of 15 m.

5) **Cone** with a radius of 10 cm and slant of 14 cm.

6) **Square pyramid** with a slant of 12 m and a base of 5m.

Find the Volume of the following figures with the given dimensions:

7) **Cube**: Side length = 15 cm.

8) **Rectangular prism**: height = 12ft; length = 6ft;   width = 4ft

9) **Sphere**: radius of 10 cm.

10) **Cylinder**: radius of 12 m and a height of 15 m.

11) **Cone** with a radius of 9 cm, slant of 14 cm and a height of 11 cm.

12) **Square pyramid** with a slant of 11m and a base of 5m, and a height of 10 cm.

You have completed the course!

Congratualtios for your hard work!

SELECTED ANSWERS

1) Integer Situations/ Evaluations 1) negative 3) Positive 5) Positive 7)   Negative 9) 20   11) 17 13) 22   15) 11
2) A) 1) < 5) < 7) >
3) 1)-10   3)-12 5)17 8)-24 17) -24
4) 1)-6   3)4   5)-9 7)6   9)4   20) -11
5) A) 1)-13   3)11 5)13 7)-26 5)B : 1)-2   3)-16   5) -10
6) 1)-10   3)-6   5)-17 7)-8.7 9)-22
7) 1)11   3)-2   5)1   7)24   9)26.6 11)-23
8) 1)-15   3)-19 5)-25 7)-13   9)-35 11)-6
9) 1)-16   3)-10 5)-14 7)22 9)1
10) 1)3   3)-1   5)-1   7)4
11) 2)0   3)2 5)4   7)-9
13) 1)12   3)-32 5)-54 7)30   9)4
14) 1)8   3)-12 5)-24 7)18   9)-48
15) 1)-64   3)-48 5)4 7)-32   9)72
16) 1)8y   3)20pq 5)8ab 7)-24y 9)20r 11)40xy
17) 1)24   3)12yz 5)8ab 7)24xy 9)3xy 11)24x
18) 1)2   3)-2   5)3 7)-6   9)1   11)-9   13)-5

19) 1)3   3)z   5)-2b 7)2x 9)x   11)4x
20) 1)-4   3)-72   5) 96 7)-18 9)7   17)30
21) 1)15 3)7 5)18 7)9 9)18 15) 14
22) 1)20 3)22 5)7 7)15 9)1 11)5
23) 1)7 3)12 5)11 7)4   9)-2 11)13 13)-13
24) 1)14 3)2 5)15   7)-4   9)22 11)7
25) 1)6   3)-3   5)7 7)-26 9)0   11)-9
26) 1)15   3)-9 5)-20 7)-19 9)22   11)-12

28) 1)5   3)-5   5)6   7)8   9) 3.5   11)-2

30) 1)10   3)-96   5)36   7)36   9)-14   11)16

33) 1)1/5   3)4/7   5)1/2   8) 2/21

35) 1)0.8   3)-1.4   5)0.5   7)-10   9)-12   11)-0.7

36) 1)-7   3)18   5)-96   7)-1/5   9)-65   11) 0

39) 1)x+11   3)20+x   5)5x-15   7)7/x   9)3×2+9   11)2x+2x

40) 1)x=7   3)13   5)x=52   7)x=3   9)x=9

41) 1)-6x+24   3)-3z+12y   5)-4a+12b   7)-6x+12y   9)-3x+3y   11)15-12x

44) 1)9x-21   3)-4y   5)-3   7)21x+6   9)x-y   11)-4

47) 1)2x   3)7x/9   5)6y/35   7)10t/7   9)9t/10

32) 1)2   3)-3   5)-4   7)-1/2   9)-1   11)-1   13)-3   15)-1   17)1   19)-1/3

49) 1) 7   3)2   5)-9   7)2   9)1

34) 1)14   3)11   5)1/3   7)-5/2   9)-7/5   11)21   13)-1   15)7/4

54) 1)12   2)56   3)125   4)16   5)24   6)4.9   7) 216

57) 3)$x \leq 5$ 4)$-2 < x \leq 3$ 5)$-2 \leq x \leq 0$ 7) x<10   9)-3<x<3

58) 1) $13x^2$ 3) $2x^3$ 5) $7y^2 + 2x^2$

59) 1)1   3)124   5)19   7) $9y^7$ 9)247   11) $3y^4$ 13) $-7y^4$ 15) $4y^3 + 19y^2$

60) 1) 3^8 3)49   5) $12^3 = 1728$ 7) y^4 9)15   11)0

61) 1) x^9 3) y^6 5)1   7) y^2 9) x^{15} 11) 81

62) 1) $1/x^3$ 5) 1   7) $1/x^3$ 9) $1/x^5$ 11)1

63) 1)81   3)64   5)256   7)12   9)648

64) 1) 4.1×10^{-4} 3) 9.1×10^{-5} 5) 2.5×10^{-3}

65) 1) 4.5×10^1 3) 4.1×10^4 5) 2.1×10^1 7) 6.1×10^5 9) 3.02×10^2 11) 2.1×10^2

66) 1) 352   3) 0.000561   5) 802   7) 0.46   9) 650000

71) 1) B(3,-3) J(-3, 0), L(0, -7), P(5,0), z(2,-2)

73) Hawa (-2,4), Jama (1,2) Aisha (3,-1)   3), Nawal (-6,5)

78) 1)A   2)E   3)B   4)C   5)F   6) D

82) 1)2   2)-4   3)4   4)0   5)-2   6)Undefined

Selected Answers –Continued :

83) 1) 3   2) 0   3) -1   4) undefined   5) 0   6) -3

84) 1) -1 3) 1 5) 2/7 7) -9/7   9) 1 11) -2/3 13) -9- 15) -3/2

85) 1) y=2x+1 5) y=5x   9) y=-3x-1

87) 1) m= 2 Y-int= 3   3) m= -4 Y-int= -2   5) m= 1 Y-int =0

88) 1) 10 3) 6.7 5) 13.2 7) 8.1

89) 1) 2.8 3) 5 5) 5 7) 12.2   9) 18.9

92) 1) 9,12   3) 34, 26   5) 112,32   7) 36, 28   9) 216, 64

93) 1) C=12.56 A= 12.56   3) C= 43.6   A= 153.9   7) False   8) C= 24.8 , A= 50.24

94) 1) 200   3) 126 5) 8/5

95) 1) 150   3) 216   5) 864   7) 1000   9) 729

96) 1) 148   3) 112   5) 120 7) 324

97) 1) 113.1   3) 1256.6   5) 201   7) 33.5   9) 524

98) 1) 207.3 3) 1885.7 5) 2212.6 7) 100.6 9) 2413.7

99) 1) 103   3) 396.25   5) 508.7   7) 33.5   9) 134

100) 1) 96   3) 88   5) 24

101) 1) 1350   3) 1500   5) $240\pi = 753.6$ 7) 3375   9) $133\pi= 419$ 11) $297\pi = 932.6$

NOTES

www.ingramcontent.com/pod-product-compliance
Lightning Source LLC
Chambersburg PA
CBHW081601220526

45468CB00010B/2722